D0775487

The Art of Lecturing

This simple and clear guide to lecturing is an example-based account of effective strategies for exciting and successful lectures for academic and business lecturers. From the lecturing mindset, combating fear and nervousness, to lecturing tricks and tactics, this book discusses a wide array of practical ideas that may surprise and help even the most experienced public speakers and lecturers. The author provides unique insights into lecturing for twenty-first century audiences, based on his academic and non-academic lecturing experiences at the University of Toronto and Stanford University; experiences which have resulted in numerous institutional, provincial, and international teaching and lecturing awards.

Further resources for this title, including lecture slides and videos of presentations and lectures, are available online from www.cambridge.org/9780521876100

PARHAM AARABI is the founder and director of the Artificial Perception Laboratory at the University of Toronto. He has won many awards for teaching, including the IEEE Mac Van Valkenburg Early Career Teaching Award, an international award given for 'inspirational classroom instruction', in 2004.

The Art of Lecturing

A Practical Guide to Successful University Lectures and Business Presentations

by

PARHAM AARABI
University of Toronto, Canada

CAMBRIDGE UNIVERSITY PRESS
Cambridge, New York, Melbourne, Madrid, Cape Town, Singapore, São Paulo, Delhi

Cambridge University Press
The Edinburgh Building, Cambridge CB2 8RU, UK

Published in the United States of America by Cambridge University Press, New York

www.cambridge.org
Information on this title: www.cambridge.org/9780521876100

© Cambridge University Press 2007

First published 2007
Reprinted 2008

Printed in the United Kingdom at the University Press, Cambridge

A catalog record for this publication is available from the British Library

ISBN 978-0-521-87610-0 hardback
ISBN 978-0-521-70352-9 paperback

To all who supported me, inspired me, stood by me, guided me, and lent me a hand when I fell down, thank you. You have always been the reason that I would get up every time that I would fall down.

To all who kicked me when I was down, misled me when I was lost, and discouraged me when I failed, thank you. You have always been the reason that once I got up after falling down, I would rush forward with an ever greater energy, motivation, and focus.

Contents

Preface

This book consists of a personal account of lecturing principles and methods which have worked well for me while teaching at Stanford University and the University of Toronto. The target audience is mainly professors who want to enhance their lecturing effectiveness, graduate students who want to jump-start their lecturing careers, as well as business professionals and politicians who are interested in learning about large-audience lecturing strategies. The figure below illustrates the main target and focus of this book.

This book is meant to be to the point and very clear in its presentation in order to be as accessible as possible. While the goal is for this to be a fun, interesting, and unique book about lecturing, it can be used as a supporting book for a short course on effective teaching and lecturing. It should be kept in mind, however, that effective lecturing is

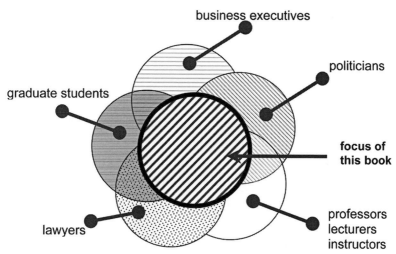

A graphical view of the potential application areas of this book.

primarily achieved by practice and experience, and not just by reading a book. It is only in conjunction with such experience that the knowledge of certain facts, methodologies, and tactics becomes useful for lecturers. Ideally, a short course on these principles, requiring approximately 10–12 hours (each hour covering a chapter of this book), would be effective at illustrating the main points contained here. Such a course should be more like a demonstration of the different issues related to lecturing (i.e. a lecture on lectures) rather than a theoretical summarization of the key points. Potentially, graduate students, academics, business professionals, politicians, and anyone else interested in giving lectures would benefit from a course based on the contents of this book.

This book and the contents therein have been made possible by the direct and indirect advice and help of numerous individuals, including Professors Jonathan Rose, Amir Keyvan Khandani, Ladan Tahvildari, Tarek Abdelrahman, Paul Chow, Sergei Dmitrevsky, Brendan Frey, Zvonko Vranesic, Lambertus Hesselink, Bernard Widrow, and Safwat Zaky. Without their contributions and help, I would still be a struggling lecturer back at Stanford University. Finally, the thousands of students whom I have had the pleasure of instructing in classes are the cornerstone of the ideas in this book. Their advice, feedback, comments, questions, and complaints were like navigation beacons that have to this day guided me through the murky waters of lecturing dos and don'ts.

This book has also been particularly strengthened by the advice and editorial comments of Pegah Aarabi and Ivana Konvalinka whose detailed examination of the book I greatly appreciate. Also, I am indebted to Sanaz Motahari-Asl for her significant help with some of the lecturing photos that appear in this book.

Finally, it should be mentioned that, unlike the plethora of books on lecturing, this book is really a first person account of the art of lecturing. Things mentioned in this book may or may not be applicable to every single lecturer. In other words, please use this book and the information in it with caution.

I Introduction

Imagine speaking to an audience of two hundred smart and highly critical individuals. If you fear them, the fear will prevent you from giving an effective lecture. If you analyze them in too much detail, then you risk confusing yourself. If you ignore them, then you will be no different than a mechanical video rerun. What you must do is to grab all two hundred audience members and bring them into your world and share with them your thoughts. You must exude confidence and remain in control of the lecture at all times. You must overcome your fear by focusing more on the audience than on yourself. Teaching, lecturing, public speaking, motivational talking, and presenting, which are all different names for exactly the same action, are an art form whose mastery can be surprisingly easy. In this book, numerous strategies, tips, and tricks will be presented that will help you with any lecturing task, including academic lectures and business presentations.

1.1 CONCLUSION

If you are interested in the fundamental ideas of this book, but do not want to spend the time and effort to read the entire book, then this section is for you. However, it is still recommended that you read this book, since much of the important details are lost in this short summarization.

Perhaps the most important lecturing advice embedded in this book is to be aware of the audience. The audiences of today are significantly affected by the presence of television and the internet in their daily lives. The shows and websites that they see, where information is packaged and spoon-fed in a careful and focused manner, result in a unique set of expectations of the lecture and the lecturer. These expectations include the requirement for an extremely organized and thought

Figure 1.1. A side view of a lecture room filled with an audience of 200.

provoking lecture (to the tune of a TV show rather than your average twentieth century lecture). So, as a lecturer, you need to think carefully about the lecture before preparing and rehearsing for it.

The mindset of the lecturer is imperative for the successful preparation and delivery of the lecture. For example, the ability of lecturers to combat their fears or to channel their emotions into positive and productive endeavors can have a huge impact on the lecture. Furthermore, lecturers who are not afraid of failing tend to deliver more exciting, passionate, and unique lectures compared with the safe and boring lectures delivered by those who constantly fear trying something new that does not work. These lecturing mindset issues should always be kept in mind long before the preparation for a lecture even begins in order to maximize the effectiveness of the lecture.

The effectiveness of a lecture is determined by three parameters: the audience quality, the lecture quality, and the lecturer quality. All of these parameters can be tuned and controlled by experienced lecturers. For example, it is important to know that the information processing ability of most audiences is akin to a narrow information channel. If you send too much or too little information, the end result will be less understanding than if you send information at the optimal rate.

It greatly helps your lecture if you offer something unique during the lecture presentation, if you connect with the audience, if you simplify and focus on the fundamental points instead of tossing out detail after detail, and finally, if you care about and are fair to the audience.

During a lecture, you need to be aware of several indicators that should ideally guide your pace and your actions. These include the noise level in the lecture room, which must be kept to a minimum, the type of questions asked during the lecture, as well as the look on the faces of the audience members. Always keep in mind that the attention span of audiences is usually far smaller than you, as the lecturer, might expect. In fact their attention span generally decays as the lecture goes on, necessitating some form of break or shock every 20–30 minutes.

Every lecturer will inevitably make mistakes. When this occurs, you must confront your mistakes directly instead of trying to hide them. Direct confrontation of errors will show a sincerity that will always be greatly appreciated by the audience. Remember that the success of the lecture is on your shoulders, hence you must be on the offensive during a lecture instead of being on the defensive (e.g. if someone keeps talking during the lecture, it is your responsibility to confront them). Breaks (every 20–30 minutes), jokes, and other fun distractions are tools that can be very effective in increasing/resetting the attention span of the audience.

The common theme in this book is to be aware of what the audience can handle and to adjust the lecture accordingly. This does not mean that lectures should be overtly simple; it just means that lectures should be focused. Packing too much into the lecture will almost always have negative consequences and result in a

Figure 1.2. A student's view of a lecture.

misconnection with the audience. In the same way, when preparing overhead slides or computer presentations, it is essential to focus on the main issues instead of producing extremely detailed and unintelligible slides.

Finally, some of the points mentioned in this book may work for you, and some may not. Always keep in mind that the rules and tactics that apply to each lecturer may differ slightly. However, the general principles in this book should allow anyone, even amateur lecturers, to give effective and understandable lectures.

1.2 A LECTURE ...

Officially, a lecture is defined as "an exposition of a given subject delivered before an audience or a class, as for the purpose of instruction."[1] In this book, the word lecture is used interchangeably for a

---

[1] *The American Heritage® Dictionary of the English Language,* Fourth Edition Copyright © 2000 by Houghton Mifflin Company.

variety of actions and sequences of events. The commonality, and the definition of the word "lecture" utilized here, is that a lecture consists of taking somewhat complex topics and ideas and delivering them in an understandable fashion to an audience. Obvious examples include the delivery of a university lecture to a large classroom of students, presenting in a corporate setting, giving a scientific presentation at a conference, or even pitching an idea for a company to venture capitalists. In all these cases there is a need to understand how to speak, what to say, and what to do during and after the lecture. This book is about these actions which are the essence of that thing that we will from now on call a *lecture*.

1.3 THE INGREDIENTS OF A LECTURE

A lecture is composed of three primary ingredients which can make or break the lecture. The first and most important element of a lecture is the audience. The audience is the reason for the lecture, the means by which the lecture can succeed or fail, and a source of energy and inspiration for the lecturer.

The next important element of a lecture is the lecturer. The lecturer is the master of the show. She or he alone defines and sets the pace, the tone, and the style of the lecture. The lecturer is essentially responsible for taking in the emotions, feelings, and energy of the audience and focusing them towards the presentation and the description of a single point.

The third and final important ingredient of a lecture is the lecturing medium. The lecturing medium is the way and form in which ideas are transferred from the lecturer to the audience. This can be thought of as a communication channel or link between the lecturer and the audience. The default and baseline modality of this channel is obviously speech. However, it also can consist of the usual PowerPoint presentations, chalkboards, slide shows, etc.

Careful and precise control of all these ingredients is often required in order to ensure a successful lecture. The audience, for example, can easily get tired, bored, or confused. This must be avoided at all costs. The lecturing medium can either be too confusing or too much (such as a crowded PowerPoint presentation); this must also be avoided. Finally, the lecturer can lose control of the audience, lecture at a pace that is inappropriate for the given medium (for example, pacing through overhead slides too quickly), or fail to maintain the interest of the audience. These points define the difference between a great and memorable lecture and a bad, boring, and confusing lecture.

1.4 BOOK MOTIVATION

A crowd of two hundred impatient students start gathering and sitting in the lecture room, anxiously awaiting your arrival. As soon as you enter the room, most eyes begin staring at you, some with fear, some with envy, but most with excitement regarding what you may do today during this specific lecture. Before the performance and show begins, you glance one last time at your notes, trying to find that one

Figure 1.3. The contest between a single lecturer and a lecture room packed full with 200 students.

lecture-killing error. Feeling confident for not finding such a mistake you erase the boards, settle down your chalks, and open your can of iced tea and have a quick drink. You then take one last look at your notes, not because you are actually paying attention to the notes but because you are nervous of what is about to come.

Seeing that you are almost ready to go, the conversations among the two hundred audience members start to die out. A quiet hum is all that can now be heard. While the noise has reached a minimum, the excitement of each of the audience members is at its maximum. At this moment, time begins to slow down, and as you pick up your first piece of chalk you start the lecture, the theatrical performance, the intellectual show.

But will you bore the two hundred students to sleep, confuse them to the brink of insanity, or energize them in a shape, way, and form that they have never experienced before? When the hour (or so) is over, and the students walk out, what will they say about you? More importantly, what will they remember about their past hour? Will they go home dreaming about the wonders of science, the possibilities of technology, the beauty of history, or the fact that they have to sit through months of boring, useless, and uneventful lectures? The answers to these questions are for the most part determined by you, the lecturer. You alone will shape the perception, the experience, and the resulting memories of the audience members who attend your lectures.

This book is not really a guide nor does it consist of foolproof methods for making your lectures exciting. It is but one person's tale of what he has learned after almost a decade of teaching, after one thousand of those moments that are described in this section, and after numerous negative and positive feedbacks from his audience members.

1.5 FROM A SHY OBSERVER TO AN UNORTHODOX LECTURER – THE STORY OF THE AUTHOR

Perhaps the best way to describe myself is by saying that I am an average man trying to have an above-average effect on the world around me. My

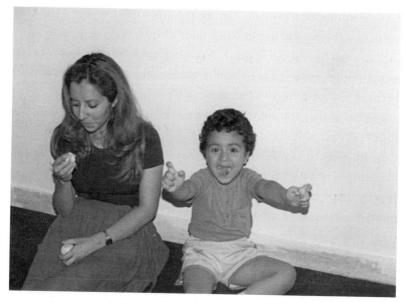

Figure 1.4. Parham, making his first attempt at lecturing, beside his mom at age 3.

story starts almost three decades ago in Iran. I was born on August 25, 1976, to an upper-middle class family in Tehran. My parents, both civil engineers, were on their way to building a financially secure and stable life when I was born. However, my birth coincided with the turmoil of the Islamic revolution, which toppled the government of the Shah of Iran.

Soon thereafter the Iran-Iraq war ignited, lasting for almost an entire decade and killing more than a million innocent people on each side. In the middle of this war, my family, who were unhappy about their lives in Iran and more importantly about the prospects of a future for me and my younger sister, decided to emigrate from Iran.

The process of emigration was long, tedious, and grueling. We first traveled to Japan with hopes of obtaining a visa to the United States. After being refused, we then moved to Switzerland the next summer, again with hopes of getting a visa for the United States. After living there for about five months, we were denied a US visa once again. On subsequent attempts, with trips to Switzerland and Turkey, we eventually obtained a jigsaw puzzle of US visas (i.e. visas for me and my sister, but not my parents, then visas for my mother and myself, and not my sister and my father, and so on). Eventually, after securing enough jigsaw puzzle US visas, we were finally able to move as a family to Atlanta, where my uncle was living.

My time in Atlanta was exhilarating and enjoyable. While I had taken English classes in Iran (from a former advisor to the Shah's family, Mr. Barzin, who advised and taught me more about life than English), I still felt shy when it came to speaking to my classmates in Atlanta. This was especially true when it came to girls in the co-ed school, whose presence in the school after coming from an all-boys school in Iran was quite an eye-opening, exciting, and nerve-racking experience. The unease with English faded rather quickly, but my shyness remained.

After the death of my grandfather and the expiration of our US visa, we returned to Iran and several months after, secured immigration to Canada. At the age of 12, I immigrated with my family to Toronto. Throughout high school, and through my subsequent undergraduate studies at the University of Toronto, I was a shy observer in most conversations and a horrible public speaker. Perhaps the largest and most frustrating problem was that I knew what I wanted to say, but when I tried to speak my shyness and nervousness would interfere with my conversations, resulting in either quiet, extremely fast, or unintelligible speech. This did improve slightly as I got older and as I obtained more public speaking experience. Nevertheless, the nervousness and discomfort with giving lectures for large audiences remained. After obtaining my Master's degree at the age of 22 from the University of Toronto, I went to Stanford to obtain my Ph.D. in Electrical Engineering.

Stanford was and still is a very unique and fantastic place. From the resort-like campus, to the friendly but tough atmosphere (which is perhaps true of a lot of places in the United States), and to the courses that are televised live on television and over the internet, my two doctoral years there were certainly ones that I will never forget. When

Figure 1.5. Parham walking down University Avenue in Palo Alto, just a few steps away from the Stanford Campus.

I went to Stanford at the age of 22, I was a shy geeky student with little or no idea about a great number of things. When I completed my doctoral studies, I was a more confident and feisty 24-year-old with a fire inside me that has since been my inspiration and motivation. The stories and events of those two years are beyond the scope of this book.

However, it is safe to say that enough interesting and unique events happened in those two years to merit an entire book! In less than two years, I became a volunteer teaching assistant, a course instructor, a soccer referee, a presidential campaign volunteer for Senator John McCain, the founder of a start-up company at the inter-section between beauty and technology, a private investigator (of sorts, not an official one!), as well as a regular research assistant and doctoral student. Throughout it all, the support of my advisor, Professor Vaughan Pratt, the meetings and support of my associate advisor, Professor Bernard Widrow, and the support of two good friends, Dr. Keyvan Mohajer and Dr. Majid Emami, were essential to get me through the rough patches.

Figure 1.6. Parham in front of his office and virtual home, which was located in the William Gates Computer Science building.

Figure 1.7. Parham in front of the Stanford Oval.

Figure 1.8. Stanford/Toronto friends Dr. Keyvan Mohajer (right), Parham (middle), and Dr. Majid Emami (left).

The support and help provided by Keyvan and Majid was tremendously important, not just during my Stanford days, but afterwards as well. It was Keyvan who observed the very first lecture of my life and gave me a great deal of positive feedback. If he had not done that, I perhaps would not have chosen this path. Majid as well has always provided me with unconditional support and help. Without Keyvan Mohajer and Majid Emami, no lecturing success, no award, no faculty position, and certainly, no book would ever have been possible.

One of the unorthodox aspects of my Stanford years that is related to my current lecturing methodology involves my preparation for the doctoral defense. In preparation for my doctoral thesis presentation, while fearing the worst given the short duration of my doctoral studies (at my request and push, the defense was scheduled 16 months after I started at Stanford, the norm was more than 60 months), I solicited the help of a well-known Stanford professor known as Bernard Widrow. Professor Widrow, or BW as we sometimes called him, was a world renowned researcher in the fields of signal processing and microphone arrays (the direct area of my dissertation). Every night for a period of 3 months, I would present my doctoral progress to BW, and he would offer feedback regarding it as well as the presentation. Getting direct feedback about how to present from a world renowned researcher has been one of the main pillars of my lecturing style. In return for BW's help, I would help him patent and research novel inventions, business ideas, and on most occasions engage in political discussions (which, by the way, were quite enjoyable given the similarity of our political views).

The preparation worked well, and I eventually defended my Ph.D. with relative ease. After teaching my first course at Stanford on a volunteer basis (by begging several Professors to give me the chance), I gained confidence as a lecturer. This confidence was essential as I started a faculty position at the University of Toronto in June 2001 at the age of 24.

The initial years as an Assistant Professor at the University of Toronto were both challenging and exciting. My teaching experience at Stanford (especially with the televised feedback) provided me

Figure 1.9. The view of a typical lecture at the University of Toronto.

with an excellent base from which to launch a lecturing career. The University of Toronto provided ample opportunity, given the large class sizes (about 100 students per class on average) and the number of courses being taught (typically 3 single-semester courses per academic year).

This setting provided me with a platform to experiment with lecturing. After five years, a few teaching awards, and after having 1500 students pass through my classes, I now realize that some of the lecturing techniques that I experimented with were quite successful. Consequently, the culmination of these experiments and what I learned as a result are the basis and motivation for this book.

2 The lecturing mindset

The mindset of an individual is perhaps the most important attribute in determining that person's success. People with positive, optimistic, and ambitious mindsets tend to achieve more than those who are negative, overly pessimistic, or lack any kind of ambition. Perhaps the most important reason to have the right mindset is to overcome and defeat fear and failure, which can only be done by optimism and a positive, resilient, and determined attitude.

Having the right mindset is even more important for certain careers where quick reflexes and constant adaptability are required. And, just like anything else in life, the right mindset for any task can be acquired easily with enough practice and training. In this chapter, we will take a closer look at the mindset required for delivering successful lectures.

Case in point
Prior to the start of my lectures, I use a few minutes to walk through the audience while engaging in short but interesting discussions with them. The conversations generally consist of short greetings and questions about their daily activities. This action, aside from improving my relations with the audience, has a very positive effect on my mindset. First, it allows me to get more familiar with the students, thereby allowing me to use this familiarity to my advantage during the delivery of the lecture. Second, it sets my mindset into a talkative state from which giving a lecture comes quite naturally. In essence, this allows for a "warm" start to the lecture rather than a "cold" start which at its worst would consist of giving a lecture without having spoken a single word prior to it. Finally, this pre-lecture audience interaction gives the audience the right mindset as well.

Figure 2.1. At the start of most lectures, a lecturer's heart starts beating faster, the excitement starts to build up, and his focus turns towards the sea of students sitting in front of him.

They perceive me as a friendly lecturer open to their concerns and questions rather than a cold and authoritative presenter. This allows the lecture to become more dynamic, interactive, and friendly.

It should be mentioned that a very important factor in acquiring the right mindset is the quality of life which we live. The usual motherly concerns of sleep, diet, physical exercise, and general health contribute significantly to how well we perform our daily tasks. Lecturing is a task that requires extensive focus and energy, and as a result, it requires all the elements of a lecturer's life to be in balance.

2.1 MOTIVATING YOURSELF

The most important thing before you undertake a complex task is to have in mind the proper motivation for performing the task. We humans, given the right motivation, can accomplish great and monumental feats. From the exploration of space, to the construction of skyscrapers, we motivate ourselves for tasks that may initially seem impossible and achieve them with the strength of our motivation.

For lecturing, having the right motivation is also very important. If the goal becomes simply to just talk for an hour and leave the room, no need will exist to try hard, to prepare, and to focus on delivering the best lecture of your life. Perhaps the best motivation is to look into the eyes of the audience and just imagine how they will look at you and what they will say about you after a fantastic, provocative, and life-changing lecture. It would also help to look at the audience and to realize how confrontational, cynical, and harsh they could be should you fail to deliver a good lecture.

Of course, for most people this motivation comes naturally since no one wants to give a bad lecture. However, my experience has shown that sometimes after a few failures, many lecturers lose their motivation and become comfortable with delivering below-par lectures. This is a common trap in life when we presume our lack of ability, talent, or luck has prevented us from successfully performing a task, a task for which we lose motivation and excitement. It is very important in life to always look for the right motivation to provide an exit from these depression traps.

2.2 FAILING YOUR WAY TOWARDS SUCCESS

Failing is an unchangeable and ever present reality of life. It is also an important part of the learning process. We humans, from the decisions of individual neurons in our brain to our collective actions, make mistakes, learn from them, adapt to avoid similar mistakes in the future, and then, we try again. Often, repeated mistakes are possible. However, at some point we have seen enough recipes for failure that we simply fail to fail.

As a result, perhaps the best way of looking at failures in life is as partial successes. From a bad lecture to a missed promotion or award that does not go your way, as long as you learn why your lecture was bad or why you did not get the promotion, and as long as you promise yourself that you will try again, the overall experience has in reality been a positive step forward. In time, we can develop an affinity and hunger towards failing by realizing that our reaction towards any failure will result in the kind of success that would otherwise have been unattainable.

Case in point

I am not sure why, but on an almost daily basis for the past year I have received requests for advice from individuals, students, and sometimes even newspaper columnists. One of the most common questions is "to what do you attribute your success?" Perhaps my first response would be "what success?" in order to ensure there is no overstatement of my accomplishments in life. However, when pressed, and after naming the usual suspects such as my family, friends, and advisors, I give the following explanation.

Perhaps one of the most important reasons for me (an average man) having a somewhat above average effect on the world around me is my reaction to failure. I seek failure. I want failure. The reason is that for me failure brings such an above average, energized, and focused response that this mere reaction provides enough fuel for successes far beyond my initial imagination! In all my life, perhaps my best kept secret, a secret that has saved me time after time, has been my focused and controlled reaction to failures.

2.3 CONFRONTING YOUR FEARS

Perhaps the greatest handicap, in life and in lectures, is our fear of what may go wrong. Fear weakens us, stops us from reaching our potential, and changes the nature of our personal daily battles from external battles to an internal civil war. As such, it is imperative that any fear be directly confronted and resolved at the earliest possible opportunity. This, of course, is a lot easier said than done!

One method of combating fear is by purposefully realizing and aiming on occasion for the worst and scariest possible outcome. This is a technique that psychiatrists often use in treating phobias. For example, a person who is scared of heights would be taken to the highest floor of a building and made to look down. Or, a person afraid of talking in a group of 5–10 people would be forced to give a talk for 200 people. Such an approach can have very negative consequences if the fear overwhelms the individual. However, if the individual is keen on eradicating any type of fear, be it for public speaking or any-thing else, this brute force sensory overload approach can be the quickest and most effective method for fear elimination.

Of course, one should only try this approach when there are no significant negative consequences. For example, for lecturers who intend to improve their lecturing effectiveness but fear trying new and possibly unsuccessful tactics, one approach would be to have an official experimental lecture where all possible tricks and tactics are utilized. The audience members should be informed about the experi-mental nature of the lecture, thereby minimizing the consequences should the lecture prove unsuccessful.

Case in point

My own nervousness in speaking to a large audience was removed by a brute force fear confrontation strategy as outlined above. I forced myself to teach a fundamental signal processing course at Stanford University, which included an audience of about 100 students and a television/web audience of approximately 1000 others. While this journey to the edge of large-audience lecturing was quite nerve rack-ing, once I got there the nervousness was gone and my fear of this nervousness was forever eradicated.

2.4 RE-CHANNELING EMOTIONS

In every human there exists a wealth of power and potential. We will probably never realize the extent of our own capabilities, except in exceptional situations. What will a mother do when her child is in danger? What will a man do for a girl that he utterly loves? What will we do when our own lives are in danger? The answers to all of these questions in some way depends on our emotional responses which have fueled the survival of our species. Our emotions are the greatest and strongest resource that we have available to tap into for accomplishing nearly impossible feats. Many of the world's most successful individuals have gotten to where they are by channeling their emotions into useful and constructive endeavors.

Some emotions cannot be useful in their raw form since they have a weakening effect on us. Fear, for example, about what may go wrong can consume us if left untreated. It can stop us from moving forward, dissolve our focus, and stall us at the worst possible moment (such as the start of a lecture for 200 audience members).

Other emotions, such as love, hate, and anger can be harnessed to give us motivation, focus, and strength. For example, anger about what is not right in our lives or what is not fair can give us the resolve we need to perform difficult tasks, such as giving a lecture for a large audience.

It is important to note that anger and fear are somewhat complementary emotions. For example, when fear stalls us, anger can be used to focus us and to motivate us to overcome and even forget about the fear. The trick in all of this is to learn to control our emotions in such a way that they become a useful and powerful resource rather than a disadvantage.

Case in point

Whenever I stand up to give a lecture for a large audience, especially if I have not done so for a long time, I get nervous. This nervousness can be best described as an innate fear of public speaking, and although in recent history I have found this initial nervousness humorous, this was not the case at the beginning of my career. Initially, when I was giving lectures for the first few times, my natural nervousness would bring about more fear, and the increase in fear would bring with it greater nervousness. This explosive cycle continued until my entire body and mind were virtually paralyzed by fear.

This continued until one day I got angry about being nervous. This anger directly conflicted with and was brought about because of my fear. The greater my fear got, the greater the anger would get, and the anger would provide the focus, the opposite Newtonian force, to

Figure 2.2. A lecturer standing in front of a large audience.

allow me to overcome the fear. In time, I learned that the relation between anger and fear is somewhat similar to the relation between water and fire. And by having control over the anger valve, I was finally free of the entanglement and bounds of my fear.

2.5 CHAPTER SUMMARY

This chapter focused on the lecturing mindset that can give a lecturer the edge in combating nervousness, overcoming fear, and recovering from failures. This mindset consists of becoming accustomed and unafraid of failure (by learning to properly and forcefully respond to it). On occasion, it might even be a good idea to seek failure in order to eradicate any fear of it. The proper lecturing mindset also consists of utilizing other more powerful emotions such as anger to combat and forget about fear.

The best lecturers are those who have a supernatural motivation for their lectures. They are unafraid of giving bad lectures (though they try their best to ensure the lecture is a success), and have a never-ending supply of energy and strength which flows from their emotions. This mindset, combined with an active, focused, and healthy lifestyle, more than anything else, is the greatest key to success for a lecturer. Without this mindset and healthy lifestyle, any trick, gimmick, or teaching strategy is just a hollow attempt that will most likely have little or no effect on improving the quality of the lecture.

2.6 Chapter checklist

- Before the lecture, try to be calm and relaxed, and just focus on the right lecturing mindset
- Motivate yourself by remembering the task at hand and its significance
- Think about the tens or hundreds of students who will need your help to learn and to grow intellectually
- Do not be afraid of failing, after all, everyone fails once in a while
- Remember that it is okay to be nervous and to have a slight fear of the lecture
- Think about your fear as an enemy, and think about strategies for confronting and eventually annihilating this enemy
- Try to re-channel your fear towards being more energized and passionate about the lecture
- If that does not work, try becoming angry about your nervousness or your fear
- Trust yourself and your abilities, be confident that you will deliver a good lecture, and then start your lecture

3 Old school basics

Advice from those with more knowledge and experience is always useful. This advice must be weighed against the fact that people and the world change, hence causing certain forms of advice to become outdated. In the case of lecturing, there are many online and printed public speaking and presenting guides that tell you the same things and give you the same lecturing pointers.

In this chapter, we will start by exploring some of the ways in which lecturing to today's internet generation audiences is different than in the past, hence causing certain classic lecturing suggestions and notions to be outdated. We will then focus on other lecturing concepts which are universally true for all audiences and all lecturing, presenting, and public speaking situations.

3.1 THE INTERNET GENERATION AUDIENCE

Fifty years ago, audiences would pack lecture rooms paying a substantial amount of attention to the speaker. Those audiences had not grown up with a television in their room, nor had they grown up with a permanent internet connection. The constant presence of television and the internet in the lives of today's youth has significantly modified the expectation that they have of a lecture.

The internet experience is almost entirely user controlled. If the user does not like a website, she/he can change the site or simply leave the computer. Furthermore, navigating through the internet is an interactive experience in that you choose when to scroll down, when to move forward on a page, and so on. While watching television is less interactive, the user still has a great deal of say, in this case by changing the channel or turning the whole thing off. Since the content providers in both media are aware of this user power, they tailor

Figure 3.1. Parham (bottom left) at his old school in Iran.

their content to be exciting and captivating, thereby reducing the possibility that the user will leave the website or change the channel.

The same content expectations exist when audience members attend a lecture. Half a century ago, such expectations did not exist and as a result lecturers could get away with being boring and dry. Today, all lecturers must be captivating and exciting. There must be a shock factor in lecturing, with surprises here and there and gimmicks to refocus the attention of the students. This book is primarily about this shock factor and the different ways that this shock can be induced to an audience.

It should be noted, however, that while gimmicks, tricks, and lecture shocks can help improve a lecture if used properly, nothing can remove the need for a clear, well thought out, and properly presented lecture.

3.2 DON'T BELIEVE EVERYTHING THEY TELL YOU!

Perhaps one of the most notable and controversial aspects of this book is the claim that certain aspects of classic lecturing no longer apply to current audiences. Many online and printed presentation guides tell you to present an organized lecture, to say what you are going to present, and to give an outline of the presentation beforehand. What they do not take into account is that outlines ruin the surprises of the lecture. If the audience knows what is coming next, everything becomes somewhat boring and dull. Would you see an outline for your favorite television show? Or, for your favorite newscast?

Most likely the answer is no. But even if there is an outline on a television show or a newscast, it is only a minimal outline about what is coming next. The outline should not be too detailed or too revealing. The best lecturers often have no outline, thereby continually surprising and shocking the audience and hence constantly grabbing their focus and attention.

In general, whenever you read or hear advice about lecturing or public speaking, always keep in mind the present reality of the audience. Today, at the dawn of the twenty-first century, this reality is that the current generation of students and even young researchers have grown up with exciting television and internet programs. Therefore, they expect the same level of excitement and energy. This might make some classic advice pointers about lecturing invalid. Of course, there is a great deal of classic advice that is still just as valid as ever. In the following sections, we will explore some of the more important classic lecturing points that are still valid today.

Case in point

My distrust and questioning of commonly believed principles stems from my earlier years and education in Iran. The rigid education system in Iran resulted in teachers and books which sometimes taught utter nonsense. As a young boy, I could not and would not accept ideas that were clearly incorrect. My resistance to such ideas only grew stronger as I realized that many people were blindly embracing these ideas.

Since those times, I have realized that blind devotion to unproven and illogical beliefs is a common human trait. Perhaps the biggest problem we humans have had in the past and continue to have is that some, due to malice or ignorance, propose illogical ideas and others, due to ignorance and lack of reason, follow those ideas. This simple problem is at the root of almost all wars, all conflicts, all religious tensions, and all of the other evil things that go on in the world.

Far less consequential, but pertinent to the focus of this book, are the illogical ideas that many have regarding public speaking and lecturing, including the need for an initial outline, or the various guidelines about lecture preparation and delivery. Lecturing is mostly an art, and no matter how many outlines are given or how many rules are followed, a bad lecture will always be a bad lecture. In order to lecture effectively, the best thing that you can do is to think, experiment, and practice yourself. Even some of the ideas proposed in this book may be inadequate for certain lecturing styles, hence any lecturing advice must usually be acknowledged as only a suggestion rather than an absolute rule.

3.3 THINK BEFORE PREPARING

The organization of the lecture, which involves the order and way in which ideas are conveyed, plays an extremely important part in the effectiveness of a lecture. This organization can only be achieved by careful and thoughtful preparation. One common error is to prepare for a lecture based on random thoughts, which can be dangerous if the random thoughts are in fact organized in an ineffective way.

One approach that can work well is to write down the main talking points during a first stage preparation, and then to perform a second more thorough organization of the talking points by taking into account the order and way that the ideas are presented. In general, it is always a very good idea to think about a lecture before making the final preparation.

3.4 PREPARE AND REHEARSE

A lecture is a theatrical performance. Just like the best actors rehearse their scenes before going on stage, so should lecturers rehearse their talk before taking the podium. The delivery of a lecture consists of numerous nuances, important moments, and fine points of detail; without rehearsing, it can be very easy to become disorganized and lost. In fact, one of the biggest mistakes made by lecturers is to become over-confident in their knowledge of the material without too much thought regarding the delivery. It is essential for amateur as well as experienced lecturers to make a detailed preparation for the lecture and to rehearse beforehand.

Preparations and rehearsals do not always require an instant play-by-play preparation of the lecture which can be time consuming and difficult. Instead, a simple pre-lecture organization of the main talking points usually suffices.

Case in point
For the courses that I teach at the University of Toronto, I prepare the night before each lecture by running through a set of major discussion points for the next day. I often include extra talking points (which I do not anticipate getting to) in order to be ready just in case my pace is a bit faster during the lecture. The talking point approach allows for sufficient organization to maximize the benefits of the lecture for the audience while at the same time only requiring a few minutes of preparation the night before.

One important aspect of preparation and rehearsal is that of problem-solving examples. Lecture examples, which are worked out live in front of the audience, must be thought out carefully and worked out prior to the lecture. A rare but costly error that lecturers can make is to discuss and solve examples whose solution has either not been worked out beforehand or the solution to which is beyond the scope of a given lecture. As a lecturer, you must remember that more than anything else, your ability to solve, understand, and explain the examples to the audience define your merit as a lecturer. Hence, every possible precaution must be taken to ensure that a successful, appropriate, and understandable solution is presented for any example posed.

3.5 CHAPTER SUMMARY

There are numerous sources of information on lecturing, including popular books, textbooks, scientific papers, how-to guides, self-help videos, and websites. While some of these sources are excellent and relevant, the majority are stuck with a twentieth century audience model. The prevalence of highly polished television programs and the internet has transformed the expectations and the learning methods of today's audiences. Such audiences can only be captivated by exciting, well organized, and carefully planned lectures. This has to be taken into account by lecturers in order to match their lecture optimally with the expectations and learning methodologies of their audience.

With all that said, there are certain aspects of lecturing and public speaking that are universal. These aspects include, among others, the need to organize and prepare before a lecture and the need to clarify the fundamental points of a lecture before preparing notes for it. Successful preparation and rehearsal of a lecture, when combined with proper knowledge about the audience and their ideal methods of comprehending information, can result in successful, clear, and sometimes memorable lectures.

3.6 Chapter checklist

- In preparing for your lecture, remember the realities of today's internet generation audiences
- Audiences have a shorter attention span and are used to the interactive nature of the internet and the polished look of television shows
- This should be kept in mind prior to and during your lecture
- Before you even start the preparation for your lecture, think about what you want to talk about
- After you have a clear idea of the lecture in your mind, you can then start the detailed planning process
- When your draft plan is complete, rehearse your lecture to iron out any errors or issues, and modify your plan accordingly

4   General lecturing principles

There are certain general principles that are very useful to keep in mind when preparing for and when giving a lecture. Perhaps the most useful things to know are the variables that affect the understanding of the audience, which include the audience quality, the lecture quality, as well as the lecturer quality. In turn, these variables are affected by sub-variables which the lecturer should always be aware of.

For example, knowing the type and information processing capacity of the audience is useful in setting the pace of the lecture. Furthermore, it is essential that each lecture offers something unique, something that the audience could not simply find on the internet or in a book. Also, connecting with the audience, simplifying and focusing on the essentials of the topic being discussed, ensuring the lack of bias prior to setting foot in the lecture room, and actually caring about each and every one of the audience members, are essential ways of improving the lecture experience and enabling the audience to connect better with the lecturer. These issues will be discussed in more detail in the following sections.

4.1   LECTURE EFFECTIVENESS

The effectiveness of a lecture can be simply and mathematically defined by the one and only equation in this book (henceforth called "The Equation"), as shown below:

$$\text{level of audience understanding} = c \times \text{lecture quality} \times \text{audience quality} \times \text{lecturer quality}$$

Where "c" is some constant of proportionality, "lecture quality" is a metric of how inherently interesting and exciting the lecture is, "audience quality" is a metric of how interested the audience is in the lecture and what their capacity to learn or attention span is, and finally, lecturer quality is a metric of how energized and/or capable the lecturer is.

The point of illustrating this very simple relationship is not to define specific metrics, but just to show the variables at play in the effectiveness of a lecture. Please note that our definition of "effective" here is directly related to the level of understanding of the audience after listening to the lecture.

Also note that implied in The Equation is the fact that if the lecture is inherently very boring, or, if the audience is not the least bit interested, or, if the lecturer is completely incompetent, then the end result will be an audience which understands very little from the lecture. Hence, in order to deliver a good and effective lecture, it is essential that all of these parameters be simultaneously maximized.

Throughout this book, we will show directly and indirectly how each of the three factors (lecture quality, audience quality, and lecturer quality) are in fact changeable and dynamic variables that with a knowledgeable and experienced lecturer can be maximized to yield a very effective lecture.

4.2 THE NARROW CHANNEL MODEL

Communication theory is a field where the optimum transmission of information via various media and communication channels is studied. Based on the theories of Claude Shannon nearly 60 years ago, communication theory tells us that for each individual communication channel (i.e. phone line, cable line, etc.), there is a maximum rate at which information can be sent from the transmitter to the receiver.

Now, consider the lecturer as a transmitter in the communication theory sense. If the transmitter/lecturer sends information at a rate that is too fast, then there will be a high probability of information loss by the receivers. If the transmitter/lecturer sends information at a slow rate, then the information is often obtained correctly by the receiver, but only a small amount of information is conveyed due to the slow rate. In all cases, there is an optimum rate where the

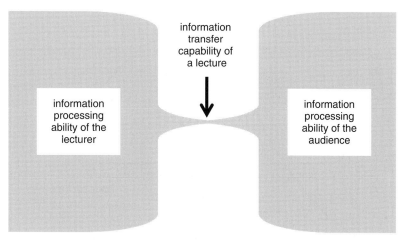

Figure 4.1. Both the audience and lecturer have significant information processing capabilities, usually (there are always exceptions to this!). However, the information channel or connection between the lecturer and the audience is typically very narrow, or at least far narrower than the amount of information that could be processed by either side. As a result care must be taken by lecturers to ensure there is no overload of this narrow communication channel/connection.

combined probability of error and rate of information transmission result in the maximum reception of information. The figure below illustrates this point.

Most lecturers make the mistake of transmitting too much information in their lectures without the slightest clue as to what the optimum rate for their audience might be. Usually, the best way to get the sense of the ideal lecturing rate is by varying the rate in the first few lectures and requesting feedback from the audience. In all cases, lecturers should be warned that the true capacity of the audience is probably much less than the capacity that is perceived by even the most seasoned lecturers.

What makes this discussion somewhat more complicated is the fact that this narrow channel model is in fact a time-varying one (i.e. it changes with time). We will get back to this point later in the book.

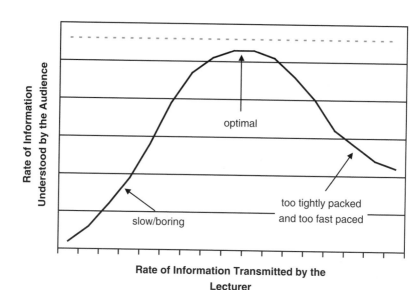

Figure 4.2. A closer view of what happens when a lecturer overloads or underloads the audience–lecturer communication channel. The ideal situation is when the lecturer tunes her/his lecturing pace with this communication channel.

4.3 BE UNIQUE

There are some aspects of audiences that do not change with passing generations or with new advancements in technology. One of the most important constant aspects of audiences is their need to hear something that is unique, unexpected, and sometimes surprising. Breaking the barriers of normality and tradition is often a simple way of captivating the audience and telling them that your lectures will be unique and refreshing.

Of course, uniqueness cannot come at the cost of reduced focus or clarity. However, it is important to note that, no matter how clear the lecture, just reading from a book or from slides will often have disappointing results.

In order to illustrate this point better, consider the typical daily television soap operas. While I have never been a great fan, the story-lines and the plot twists are often comically outrageous. The reason for the bizarre plots is to captivate their audiences on a daily basis. As a

Figure 4.3. A lecturer engaging the audience by walking close to them.

lecturer, while you do not need the same level of outrageousness, it is important to note the need and desire of all audiences to be refreshed and surprised during the lecture.

This can of course be done in many different ways. Some of these methods are explained later in this book. However, the uniqueness issue needs to be considered right at the point of preparing a lecture. A good place to insert unique elements is in the examples discussed during the lecture.

Case in point

In a third year probability course that I used to teach, most of the book examples discussed simple "colored balls in a bin" or "coin toss" scenarios. After a few lectures, it became apparent that the audience was being bored to sleep by these childish examples. As a result, I started to embed unique and interesting examples in each lecture to provoke, shock, and engage the students, including examples such as "what is the probability of a girl in class liking a specific boy?" or "how many aliens were hiding in Area 51?" The concepts behind these unique examples were the same as those in the book, but their unique storylines went a long way in engaging and interesting the students.

4.4 BE HONEST

If you have ever watched the popular news channels, you know how hard it is to find a politician who is 100% honest or one who does not spin issues in their own favor. The art of spinning has become a fact of life in many facets of our everyday lives. Politicians are most commonly associated with spinning facts and utilizing false excuses to disassociate themselves from a wrong deed or an incorrect decision. This has spread throughout society from large corporations to infamous celebrities. People, including ones who will attend your talks, are bombarded by this spin, and they are for the most part tired of it.

This spin has resulted in a distrust of authority figures, including lecturers. Such distrust results in a set of negative expectations that the audience typically has of any lecturer. For example, some among the audience may expect you to hide your lack of knowledge about a certain question. They could also expect you to hide errors or to blame them on someone or something else. In essence, they expect you to spin and excuse any of your less-than-perfect actions.

Honesty, especially self-deprecating honesty, can be used to great effect in such situations. For politicians, business executives, or lecturers, honesty can have a tremendous impact on the audience and their belief in you. If used correctly, it will be something that they do not expect, and it is something that in today's age of spin they are hungry for. For example, one look at today's most popular politicians will reveal that those who are (or at least are portrayed to be) honest, tend to have much more favorable ratings and followings. Such examples include US politicians Barak Obama, and my personal hero, John McCain. For a lecturer such honesty includes being truthful about what you know and do not know, what you believe is 100% fact and what you believe is uncertain, as well as other taboo topics such as final course average targets. Being honest also includes directly confronting any mistakes that are made during the course of the lecture, which is a topic that will be discussed later on in the book.

4.5 BE AWARE OF YOUR AUDIENCE

It obviously goes without saying that the style of a lecture should depend on the type and maturity of the audience. Consequently, knowing your audience and their capabilities is an important factor in the lecture preparation process.

Generally, undergraduate university students in their early 20s tend to be the most ideal lecturing demographic since they are mature enough to have a minimal (though occasional) amount of discipline while at the same time being open to new lecturing styles and tactics. Lecturing to younger students, who are still susceptible to unique lecturing styles, tends to be more difficult due to their lack of discipline. On the other hand, the most difficult group to lecture to is older

Figure 4.4. A close-up view of a typical audience.

and more mature adults (i.e. business professionals, professors, lawyers, etc.). While this group can be equally influenced by the style and mannerisms of the lecturer, they tend to be far more skeptical and tougher on the lecturer. It is the lecturer's responsibility when giving a talk in front of this demographic to appear intimidating and appropriately confident, while at the same time displaying a friendly, welcoming, and caring personality.

One strategy that might be helpful with tougher audiences is the Widrow Maneuver, named after Stanford University Professor Bernard Widrow. The Widrow Maneuver consists of giving a very understandable, simplified, focused, and clear talk to a mature audience. However, once in a while, he would flash a page of mathematical proofs (i.e. very complex and impressive equations) related to one of the understandable points in the lecture. This page of complex equations would not be discussed in detail by him during the lecture (though he would say a few words about what the page contained) but he occasionally ended up answering questions about them at the end of the lecture. This Widrow Maneuver can be analogized as showing your opponents (the audience) your guns (full page of complex equations, figures, etc.) without actually firing!

4.6 CONNECT WITH THE AUDIENCE

One of the most important factors in determining the quality of the lecturer is her/his degree of connection with the audience. Most successful lecturers essentially "warm up the crowd" before and after the lecture by talking to individual audience members and/or chatting briefly with a subset of the audience. During the lecture, looking directly at individual audience members, addressing them using their names (which could be learnt during the pre-lecture discussions), and personalizing the lecture are methods of establishing a strong connection with the audience.

One may wonder how successful this "name learning" business will be in a lecture with 200 audience members. As it turns out, even in large lectures, remembering the names of just a few individuals

Figure 4.5. A lecturer addressing individual audience members in an attempt to better connect with them.

can go a long way towards increasing the level of personalization of the lecture. Care should be taken, however, because getting too many names wrong will just make the lecturer look like the "nutty professor!" One successful method of ensuring that the names are

Figure 4.6. A "warmed-up" audience who are mostly responding positively to the lecture and the lecturer.

learnt correctly is to ask the names of the audience members who tend to talk a lot or ask many questions. This way, by narrowing down the total list of names to be remembered, the probability of a name association mistake is reduced.

Another and perhaps easier way to build a positive lecturer–audience connection is to engage the audience before or during the lecture. By asking them audience-centric questions (such as "what is your opinion on this topic?" or "you look tired ... tough day?"), you will be directly increasing their level of comfort with you, and in the long run, their level of comfort with the overall lecture.

4.7 SIMPLIFY AND FOCUS

The worst lectures tend to be those that have no focus, no point of convergence, and no basic themes that can be stated in a simple and understandable manner. The process of simplification and focusing of a lecture is often useful for the lecturer in order to understand

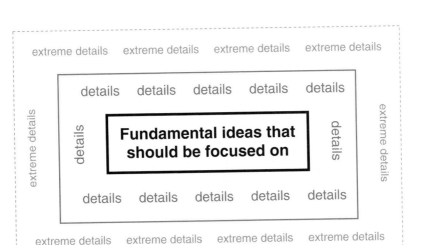

Figure 4.7. There will always be details and extreme details. However, the goal of any lecturer should be to strip away the unnecessary or somewhat necessary details in order to focus on the most fundamental concepts.

the essence of what does and does not need to be discussed. Lectures that are built up from a focused and simplified point tend to be far more organized and understandable than lectures that are a cluttered sequence of partially related points.

One thing that is often ignored is that audience members are not automatons who lack the ability to learn by themselves. They do not need to hear about every aspect of a certain topic. In fact, the job of a lecturer is NOT to discuss a subject in its entirety, but to raise the curiosity and interest of the audience while making them understand the fundamentals of a subject. As a result, when preparing for lectures, you must decide what is fundamental and essential to be included in your lecture, and what is not as essential that the audience can learn/explore/study by themselves without extensive coverage during the lecture. Most lecturers tend to overestimate the number of fundamental/essential lecture topics. Consequently, always consider and reconsider your talking points to ensure consistency and flow. In cases where a topic is important but not fundamental, it is useful to point the audience briefly in the right direction for future exploration.

It should be noted that the discussions in this section do not mean that lectures must always be oversimplified for all audiences. In many cases it makes sense to give lectures that are slightly faster and more aggressive than what the audience is used to, after all, that is how audiences grow and mature academically. Nevertheless, even in those cases it helps to have focus and clarity during the lecture in order to maximize the audience's level of understanding.

4.8 REMOVE ANY AND ALL BIASES

Within our multicultural society, it is very likely that today's audiences will be from all over the world and will be affiliated to different religions, ethnic groups, sexes, and beliefs. As a lecturer (and just as importantly, as a teacher), it is fundamental that you perceive and treat all audience members fairly and equally irrespective of their beliefs, ethnicity, sex, or their mastery of English. There is really nothing more that can be said about this point, except the following case in point which you may find interesting and useful.

Case in point
There is an interesting anecdote that has guided my perception of people for the past 20 years. When I was 9 years old and living in Iran, the Iran-Iraq war was being waged with full force, killing thousands of young and innocent people daily. With all the bloodshed, the front lines changed and moved very little (in fact, by the end of the war the border lines remained almost exactly where they were before the war). One of the larger changes during the war, however, was when Iraqi forces under Saddam Hussein took over the city of Mehran.

With a massive Iranian response, and after numerous casualties on both sides, Iran took back Mehran. Afterwards, the bodies of dead Iraqi soldiers, most of whom were close to my age, were paraded on Iranian national television. I remember that as my family and I watched this, some of the men in the family expressed their pride in Iran's victory (akin to winning a national soccer game). At that

moment, as a 9-year-old, I also started feeling proud, only to be awak-
ened by the sound of a woman crying. I turned around and noticed my
mother crying at the images, expressing that these Iraqis were just
children with hopes and dreams for the future just like the hopes and
dreams of her son (i.e. me).

The moral lesson that I learned on that day, one which has
always stayed with me, is that all people, irrespective of their ethnicity, religion, sex, or any other criteria for dividing people, are people
with hopes and dreams for the future just like us. Whenever I step
into a lecture room for the courses I teach, I do not see the mixture
of ethnicities or sexes or races. All I see is a room full of "hopes
and dreams"; hopes and dreams that can be effectively coalesced and
utilized during the lecture!

4.9 CARE

It is always imperative for a lecturer to keep in mind the humanity of
the audience. Unfortunately, it is easy, especially when teaching large
classes and lecturing in large lecture rooms, to lose sight of this fact.
It is also easy to treat the audience members and their questions
as unimportant, unintelligent, or irrelevant. Nevertheless, it is the
lecturer's burden and responsibility to develop a relationship with the
audience that portrays care, interest, and attention.

Caring implies listening to even the strangest questions and
showing respect. Caring also consists of being fair when it comes to
assigning grades, grading or re-grading exams and tests, and being
helpful when students come for help. All of these elements, which
contribute to the stature of the lecturer, play an important role in
constructing a positive lecturer–audience relationship.

Case in point

Caring about students and audiences usually goes hand-in-hand with
being a good lecturer. I have observed this in numerous lecturers, both
good and bad. Lecturers who are impatient and uncaring get annoyed
at being in a classroom dealing with people that they do not care about.

Consequently, they put little effort into their lectures resulting in a poor lecture. On the other hand, caring lecturers spend time and effort on their lectures and on their audiences, thereby resulting in better planned, more energetic, and more enthusiastic lectures.

One of the people from whom I have learned the most about being a "good" person and caring about others has been my colleague Professor Jonathan Rose. Professor Rose is one of the most caring and ethical individuals that I have ever encountered. He also happens to be an exceptional lecturer. I have often had the privilege of observing Professor Rose's departmental student recruitment lectures and have had numerous conversations with him on a variety of issues. The care that he has for others shines through in normal conversation and during his lectures. Just like the students in his classes, I have always found his genuine caring and moral clarity extremely refreshing and inspiring.

4.10 UNDERSTAND BEFORE YOU LECTURE

Perhaps the most important advice for a lecturer is to ensure complete and full understanding of the subject matter prior to the lecture. During the lecture, issues, questions, concerns, errors, and distractions will arise that will make the delivery of the lecture difficult and confusing. Unless the lecturer has a very deep understanding of the subject and all of its extreme details, difficult and embarrassing situations will arise such as a question whose answer the lecturer does not know or a proof for a certain theorem which confuses the lecturer.

With that said, the first time lecturing on a particular subject will very likely result in a few such moments. Usually by the second or third iteration you will have seen enough questions and distractions to enable you to be comfortable with almost any unexpected event.

It should be noted that in the event that you do not know or are confused about a part of the lecture, it is important not to sugarcoat the situation but to be very honest with the audience. Attempts at covering up in these situations will inevitably result in a much greater disaster than just admitting that you have forgotten or are unsure about something.

As an interesting side note, there is an old saying that the best way to learn something is to teach/lecture on the subject. This is, in my experience, absolutely true. By lecturing on a topic you are forced to learn every nuance and every aspect of the subject. Hence, your understanding becomes far greater and deeper than if you had just read a book or heard a lecture on the topic.

4.11 CHAPTER SUMMARY

If there is a single point to be taken away from this book, it is that the success of a lecture, as defined by The Equation, depends on the quality of the lecture (including the way it is produced, the style and detail of topics covered, and the overall organization of the lecture), the type and style of the audience (including how tired or bored they may naturally be, how tuned to the lecture topic they may be, and how well their personalities match the lecture style), and of course the experience, pizzazz, and talent of the lecturer. All of these factors can be adjusted by the lecturer to result in the most effective lecture for any given audience.

In order for these factors to be positively adjusted, certain points need to be kept in mind. First, the audience's ability to comprehend information is similar to a narrow communication channel which can accept far less information than what most lecturers believe. Just like a communication channel, sending too much information or too little information will result in a suboptimal transmission rate. It is only by transmitting at the appropriate rate, matched to the channel, that the maximum amount of information can be conveyed.

Next, it is important for a lecture to offer something unique to the audience. Everyone can read a book or read through a website. Why should people spend their time attending your lecture? What is unique, special, and attractive about the lecture that will captivate the audience and give them a valuable return on their time investment? These issues must be continuously kept in mind by any lecturer.

Other points to keep in mind include the need to know the type and style of the audience that will be lectured to, to simplify and focus the lecture down to the most important points (instead of covering every minute detail possible) and to connect with the audience through fairness and caring. These are some of the fundamental issues that allow an average lecturer to give an outstanding and memorable lecture.

4.12 Chapter checklist

- The effectiveness of your lecture depends on your ability and performance as a lecturer, the attentiveness of the audience, as well as the quality and content of your lecture
- These parameters should be kept in mind during the lecture, as well as during the planning stages of the lecture
- In a lecture, the audience and the lecturer are interconnected through a narrow communication channel
- If you lecture too quickly, you overload this channel and as a result the audience will not be able to process the ideas from your lecture
- If you lecture too slowly, you are not utilizing the full potential of the audience
- When lecturing, remember to be unique – uniqueness intrigues and excites the audience, making them more focused on your lecture
- Be honest in any lecturing situation – audiences pay far more attention to an honest lecturer than to one who is perceived to be dishonest or deceptive
- Know your audience and know the type of audience you have – this is the only way to adjust your lecture to their specific capabilities, needs, and wants
- Try to connect with the audience by talking and interacting with them before, during, and after the lecture – this connection results in a very positive lecturing environment that makes your job as a lecturer easier
- Make sure your lecturing points are as simple and as focused as possible – avoid rambling on or getting sidetracked at all costs

- Remove any biases that you might have personally – all your audience members are and must be treated as equal and valued participants in the lecture
- Care about your audience – if you do not or cannot, you should not be setting foot in the lecture room
- Make sure you understand the lecture before you lecture – there is nothing worse than a confused or incompetent lecturer making a fool of himself in front of an audience of 500

5 At the beginning of the lecture

The start of a lecture is perhaps the most exciting and at the same time the most nerve racking moment of a lecture. This is when the audience's mentality is changing from a normal mode to a listening and paying attention mode. This is also when the lecturer is uncertain about the lecture, the questions that may arise, and the reaction of the audience.

At the same time, the initial few minutes of a lecture define the pace, extent of success, and the overall effect of a lecture. A bad start to a lecture makes the delivery of a successful lecture extremely difficult. On the other hand, a good and energetic start can make the delivery of a smooth lecture almost automatic. As a result, a great deal of care and attention is required by the lecturer in order to ensure that the lecture starts out in the right way. In this chapter, the dynamics and issues of the first few minutes of a lecture will be analyzed in detail.

5.1 THE INITIAL LACK OF ATTENTION

The first and most important aspect of the beginning of a lecture is the attention span of the audience. While generally the attention span of the audience decreases as the lecture goes on, in the first few minutes of a lecture the attention span actually increases from an initial low point, as shown in the following figure.

At the beginning of a lecture, the audience is initially in a walking, talking, conversing, or otherwise-thinking mode. We can collectively call this the normal mode. The transition from this initial normal mode to one of understanding and listening to a lecture is neither immediate, nor easy. Listening attentively and understanding somewhat complicated concepts requires a significantly greater amount of focus and energy than usual. As a result, the initial level of attention of the

Figure 5.1. An audience at the beginning of a lecture.

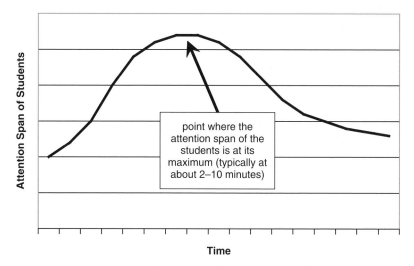

Time

Figure 5.2. This graph illustrates the average attention span of a typical audience during the first few minutes of a lecture. Initially, as the lecture starts, the audience requires time to switch to an understanding/listening mode. Hence, the initial level of attention is low. As the lecture goes on, the mode of operation of the audience begins to change, resulting in an increase in the average attention span. This increase continues to the maximum point at which point fatigue and boredom start reducing the level of attention of the audience.

audience is low. As the lecture goes on, the mode of operation of the audience begins to change, resulting in an increase in the average level of attention. This increase continues until the audience has fully transitioned to a lecture listening mode, which corresponds to a maximum in the attention span graph. At this point, fatigue and boredom start reducing the level of attention of the audience.

While we will discuss and analyze the effects of boredom and fatigue later in this book, it is important to note that the initial portion of a lecture is similar to the start of a car race or a horse race. Momentum needs to be built in order for a car or a racehorse to reach their maximum speed. Similarly, time is needed in order for the audience to transition fully to their lecture listening/understanding mode. However, just as a race car driver may press the gas pedal to the maximum limit during the start of the race, a lecturer should also push strongly and forcefully forward at the beginning of the lecture in order to achieve the highest possible momentum during the initial moments of the lecture.

5.2 START RUNNING FROM THE GATE

Imagine you are about to start a race, a race where it takes a few minutes for you to reach your full speed, and, a race in which fatigue will cause you to slow down after a while. What do you do coming out of the gate? Do you run out as fast as possible, or, do you slow down with the hope of delaying the onset of fatigue?

The best answer is obviously to go as fast as possible without inducing permanent damage. For a lecture, the ideal start can be described in a very similar way. It is best to have a heavily loaded and fast paced beginning to the lecture in order to maximize the mode switching ability of the audience. In the first few minutes of a lecture, when the audience is adapting to the lecture pace and difficulty, forcing them to adapt to something that is faster and heavier than what you want to maintain can be advantageous.

The beginning of a lecture often contains either a recap of a previous lecture or the easy and redundant parts of a new topic. A fast and heavy pace at this point in the lecture can keep the students focused while at the same time getting them ready for a fast and heavy remainder of the lecture. When the lecture unfolds and the pace turns out to be slower than what the audience expected, the result is a significant mental and perceptual advantage for the lecturer.

Case in point
During my lectures I use the first 3–5 minutes to go through the initial background and the simple basics of the topic under discussion. I do so at a fast but manageable pace which gradually decreases as the material becomes more complicated and as the audience becomes more fatigued. Decreasing the pace of the lecture, I have found, is always an easy and welcome action whereas increasing the pace, especially as the audience is becoming more tired, almost always has negative consequences.

    Another benefit of this initial "burst mode" lecturing style is that it wakes up the audience and forces them to focus on the lecture. If given enough time, the audience could take 10–15 minutes to catch up with their neighbors, check their phone messages, think about their day, etc. An initial burst to the lecture makes them forget about all of that and just puts one idea (i.e. the lecture), one issue (i.e. the topic being discussed), and one image (of the lecture) in their mind.

5.3   THE 5 MINUTE RULE
At the start of a lecture, there is a short period of time during which the audience tries to get a sense of the lecture and the lecturer. During these initial 5 or so minutes, the audience will determine the degree of success that a lecturer can achieve. Even if the audience has seen a lecturer presenting before, the first 5 minutes of a lecture is still an important time during which they will reevaluate the lecturer. At the end of this time, either the lecturer has made a good impression on the audience, at which point successful delivery of the lecture and rallying the audience behind the lecturer become very probable, or, the lecturer fails to make a good impression at which point almost all hope of success is lost for the remainder of the lecture.

There are several important points that need to be conveyed in the initial portion of a lecture. First, a lecturer needs to exude a significant amount of confidence and knowledge concerning the subject under discussion. This can be done by carefully rehearsing the initial moments of a lecture in order to deliver a crisp, clear, and perfectly delivered lecture. Second, the lecturer must show care, concern, and attention towards the audience in order to enable them to become an important part of the lecture rather than non-participating observers. This can be done by asking them questions and requesting feedback from them about the lecture. Finally, a lecturer needs to show enthusiasm and excitement regarding the lecture. If the lecture is delivered in a dry and boring fashion, even if done so in a clear way, it will motivate the audience to view the topic as dry and boring, hence turning them away from the lecture.

These lecturing points must be stamped in the hearts and minds of the audience in the first few minutes of the lecture. By doing this, the audience will be attracted to the lecture to the point of listening and understanding out of interest and excitement instead of being forced to do so. As a result, the next time you step into a lecture hall or a presentation room, remember, the first minutes count more than any other and first impressions definitely do matter when it comes to lecturing.

Case in point
Knowing the importance of the first 5 minutes of a lecture, I always rehearse the start and the initial sections of my lectures. Even on occasions where I have little time for a thorough rehearsal, I still ensure that the first 5 minutes have been very carefully planned and well rehearsed. On the occasions that I get these first few minutes of the lecture right, the rest of the lecture flows easily and naturally, even if the remainder of the lecture has not been rehearsed or planned carefully. On the other hand, a botched start to the lecture usually ruins the entire lecture no matter how hard I try to overcome the initial failure.

Of course, those first 5 minutes of the lecture are the most difficult for another reason as well. When the lecture starts, the lecturer is usually at her/his lowest moment in terms of lecturing ability. For example, at the start of my lectures I find that the combination of the initial nervousness along with an unprepared vocal tract makes the delivery of the lecture difficult. It takes a great deal of energy to overcome these initial problems quickly. Regardless, if I have not overcome these problems within just a few minutes, I usually end up delivering a poorer lecture than I would otherwise.

5.4 OVER PERFORMING IS BETTER THAN UNDER PERFORMING, AT LEAST INITIALLY

If under performing (i.e. being too boring, slow, or monotone) during the first few minutes of a lecture is a bad thing, then significantly over performing (as compared to usual) is also not ideal. Confusing or substantially scaring the audience during the initial portion of a lecture has obvious negative consequences. It turns the audience away from the lecture, reduces their motivation and attraction to the subject, and gives them less incentive to listen to the remainder of the lecture. Overwhelming and scaring the audience away from the lecture should never be tolerated at any point in the lecture, especially at the beginning.

However, as long as confusing the audience is avoided, slightly over performing (lecturing at a pace that is faster, more intense, and at a level that is greater than usual) or very slightly scaring the audience may actually be a good thing. In essence, if the type or capability of the audience is unknown, lecturing more heavily initially and then gradually reducing the pace of the lecture is far better than lecturing too slowly and increasing the pace as the lecture progresses.

Case in point

It is easy to discuss lecturing theory while sitting comfortably at home or at the office. In practice, lecturing is dynamic, nerve racking, and unpredictable. Hence, the precise optimization of any parameter is often hard if not impossible to do in real-time during the lecture. For example, perfectly tuning the initial pace of the lecture to ensure that it is neither too fast and intense nor too slow and boring is a hard feat to accomplish, especially when an audience of 500 keeps staring right at you!

However, I have found that by erring on the side of intense and fast paced lectures (to a slight but definitely not extreme extent), often a practical and successful solution can be achieved. Fast paced lectures are generally easy to deliver since they turn the focus of the lecturer on the dynamic and rapid evolution of the lecture instead of on their own nervousness. In contrast, slow and boring lectures will do little in distracting the lecturer while making the audience addicted to a slow and shallow lecturing pace.

5.5 THE REVIEW

Some lecturers prefer to do an initial review at the beginning of each lecture. This review would consist of a quick summary of the important points that were either discussed previously or are required knowledge for the current lecture. While such reviews are advisable in situations where a single topic spans multiple lectures, nonessential reviews are usually not advised since they overload the audience unnecessarily.

A good methodology for deciding for or against the inclusion of a review at the beginning of a lecture can be found in the realm of television shows. Shows that span multiple episodes often include an initial recap in each of the follow-up episodes. As a result, the viewer who just tunes in for the current episode will have a high level understanding of the events that have transpired in previous episodes. As an added benefit, loyal viewers who have watched all the previous episodes also benefit from this refreshing of their memories.

Similarly, in a lecture, audience members who have either not attended the previous lectures or have forgotten their contents need to be able to follow the lecture by at least having a partial high-level understanding. If a review can significantly help in this regard, then it is useful and should be included.

However, reviews should always be brief and to the point. Reviews that take almost as long as covering the original material are not really reviews, but a repetition of prior lectures. Although a full lecture repetition after a failed lecture is a good idea, excessive repetition disguised as a review will bore the audience, confuse the focus of the current lecture, and take valuable time away from the lecture. Hence, reviews are only recommended in special situations and only if the review is brief and to the point.

5.6 CHAPTER SUMMARY

This chapter analyzed the importance of the first few minutes of a lecture. It was suggested that in the first few minutes, the right impression needs to be set in the minds of the audience. Furthermore, the audience members need to be significantly motivated and even slightly scared initially in order for them to be shocked into listening and understanding for the remainder of the lecture. Of course, this shock needs to be administrated in moderation in order to not confuse and scare the audience beyond repair.

These actions need to be taken to counteract the initial lack of attention by the audience as well as the initial nervousness of the lecturer. Initially, as the lecture starts, the audience members start to focus more and more on the lecture. The more a lecturer motivates, inspires, and (to a certain moderate extent) scares them, the greater the amount of attention that they will pay for both the initial few minutes and also for the remainder of the lecture.

5.7 Chapter checklist

- When a lecture starts the audience initially pays little attention to it
- As the audience members begin to focus on the lecture, their attention levels increase, reaching a maximum at about the 2–10 minute point after the start

- The height of this maximum is determined by you, so when starting, make sure to shock the audience by running fast and hard out of the gate
- A fast and heavy initiation to the lecture forces the audience to adapt to an above-average pace, which allows you to reward them in the remainder of the lecture by lowering the pace back to normal
- This initial fast pace must of course be within reason – going extremely fast will cause the audience to be lost right from the outset of the lecture
- Remember that you have at most 5 minutes to make your impression on the audience – a bad initial 5 minutes will make delivering a good lecture very difficult
- If and only if it helps, start your lecture with a review of the necessary background material

6 Things you should be aware of during the lecture

The job of the lecturer is to teach a concept or an idea to an audience. However, many lecturers often forget this and instead focus on just the lecture instead of the audience (to the point that some lecturers do not even look at the audience). Needless to say, focusing on just the lecture often makes the lecture itself a failure. The audience's understanding of a topic is the goal. To maximize or optimize this level of understanding, it is very important to routinely observe the audience, gauge their level of interest and/or connection to the lecture, and to adjust the lecture accordingly. In this chapter, we will focus on the very important mechanism of observing the level of interest of the audience as well as on the process of correctly interpreting visual and verbal feedback.

6.1 NOISE LEVELS

As a lecturer, your job and duty is to maximize the quality of the learning environment for the audience. One important factor that controls the quality of a lecture is the noise level in the room. If you have ever sat in the back of a noisy lecture hall, you will know how hard it is to understand the lecturer amongst all the noise. As a result, noise anywhere among the audience is something that should not be tolerated in any way.

Noise almost always starts with just a quiet conversation between a few audience members. If no action is taken in response to this initial conversation, the rest of the audience will notice, resulting in other sporadic conversations. After a while, the noise wave propagates in the lecture room and the quality of the lecture quickly diminishes. The most astute lecturers listen for the relatively quiet conversations during the pauses in their lectures. Once detected, a

Figure 6.1. A "Where's Waldo" example of a lecture. Try to find the several students talking, or the students sleeping, or even the one student sleeping behind a schoolbag.

humorous comment or simply walking up and chatting with the noise-makers usually solves the problem. If the initial disruption goes unnoticed and the noise spreads, however, controlling the room and the audience becomes much more difficult. In such an event, a complete halt to the lecture is often necessary in order to restart and refocus the lecture and the audience.

Usually, the level of interest of an audience is gauged closely by the number of random conversations in a room. The more the conversations, the less interested is the audience. As a result, this noise level can also be used as an effectiveness indicator for the lecture.

The worst possible course of action for a lecturer is inaction when noise has spread throughout the room. These lectures either have already lost or will quickly lose the interest of the audience. Hence, the moral of the story for all lecturers is always to listen to the audience and take immediate action at the first sign of trouble.

6.2 VERBAL FEEDBACK

Noise levels during a lecture are one sign of the degree of interest of the audience towards the lecture. A more direct way to gauge the effectiveness of a lecture is to analyze the type and style of questions that are asked. Verbal feedback, which is really the types of comments and questions that the students raise, can be a good indicator of the level of audience comprehension. Intelligent questions and comments that are in-line with the topic of the lecture would indicate that at least some in the audience have followed and understood the lecture. On the other hand, questions that seem out of context or unnecessarily simple are often an indication of a lecture–audience mismatch. As a result, care must be taken when handling and responding to student questions in order to ensure that the lecture does not outpace the comprehension ability of the audience.

With all that said, it should be noted that there are always exceptions to every rule, as illustrated in the following case in point.

Case in point

In one of the classes that I taught several years ago to second year Engineering Science students at the University of Toronto, there was one student who clearly stood out from his peers. During every class, he would repeatedly ask questions about the most mundane aspects of the lecture. He would sometimes even ask about irrelevant issues completely unrelated to the lecture or topic. Interestingly, on rare occasions a few of his many questions would turn out to be deep and insightful. This bizarre circumstance that a rather intelligent student would continuously ask unnecessary questions was a dilemma for my interactive and open teaching style. Even groans and angry remarks from other students could not stop the questioning student.

My initial reaction to him was to attempt to answer any and all questions. This quickly became impossible given the number and types of questions. My next method was to answer his intelligent questions, and quickly pass on the unrelated questions. This cat-and-mouse game continued for the remainder of the semester, with some variations on my "niceness" towards him depending on my mood and the mood of other students.

The one thing that I learned that semester was that student questions may not always be a fair indication of the status of the entire audience. Hence, as lecturers, we need to focus on a variety of percepts, only one of which is verbal feedback.

6.3 VISUAL FEEDBACK

An often underutilized indicator of the status of an audience is the look on their faces. It is a good idea for lecturers to occasionally glance at the audience, take notice of what percentage look tired or are even sleeping, and to take action if needed. Other clues include the visual focus of the students (i.e. are they looking at a newspaper, each other, or the lecturer). One of the first signs of a bored audience is the lack of visual focus and tiredness/relaxation of the facial muscles.

Figure 6.2. A close look at the faces of an engaged audience. The visual feel of any lecture which consists of the individual looks on the faces of the students, can supply a great deal of information to the lecturer.

As in the previous sections, once you determine that the audience is bored and has lost touch with the lecture, it is very advisable to either halt the lecture temporarily or to at least attempt to change gear by injecting humor and/or interesting stories into the lecture. Either way, the lecturer needs to wake the audience up and focus their attention. The worst mistake that can sometimes be made by lecturers is to continue with the lecture at the same pace and on the same topic even if a large number of people in the audience are asleep.

One note of caution is that the fatigue and focus of the audience will vary from day to day and from lecture to lecture. Depending on their state of mind and history, it may be possible for certain audience members to fall asleep even if the lecture is exciting. Nevertheless, even in such cases, you as a lecturer must take note and attempt to ensure the focus of all audience members is on the lecture. In other words, though it may not be your fault, it is your responsibility.

6.4 ATTENTION SPAN OF THE TYPICAL AUDIENCE

It is often important to keep track of the amount of attention paid by the audience to the lecture. In general, this level of attention is a time varying one, and while it differs for different audiences, some generalizations can be made. The average attention span of an audience (representative of the audience quality) can be roughly modeled as shown in the following figure.

Initially, the attention span of the audience increases at the beginning of the lecture as everyone gets settled and focused on the lecture. As the lecture progresses, the attention and focus of the audience decreases. After some time, say T_{max}, only a minimal amount of information is passed on to the audience (since the amount of information gained or understood by the audience is directly proportional to the audience quality, as defined by The Equation). The amount of information passed on and understood by the audience, which is proportional to the area under the graph in the figure above, is more or less the same whether the lecture time is infinite or if the lecture ends at time T_{max}.

Figure 6.3. An audience on its way towards a reduced attention span.

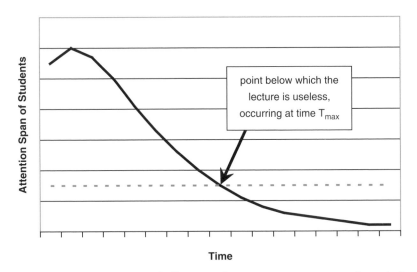

point below which the lecture is useless, occurring at time T_{max}

Time

Figure 6.4. This graph illustrates the average attention span of a typical audience. Initially, as the audience members get more excited and set regarding the lecture, their attention span or ability to understand new concepts increases up to the maximum level for the lecture. After this maximum, boredom and fatigue will weigh on the minds of the audience resulting in an exponential decrease in their attention span. After some time T_{max}, a point will be reached when the continuation of the lecture is no longer useful.

While the value of T_{max} can change based on the maturity and type of audience, for most audiences it tends to range from about 10 minutes to 40 minutes, with an average of 25 minutes. Later on in this book we will come back to this attention span question and see how the area under the graph above can be artificially increased.

6.5 CHAPTER SUMMARY

The attention span of any audience usually decays as the lecture progresses. The more time the audience spends listening to the lecture, the more likely it is that they will be tired and out of touch with the lecturer. This chapter focused on the types of metrics and feedback that can be used to estimate the attention span and state of the audience. This feedback includes visual feedback, verbal feedback, and acoustic/noise feedback. Visual feedback, consisting of the look on the faces of the audience, can be a telling sign of fatigue and boredom. If signs of fatigue are observed on a large portion of the audience, then there is a clear indication of a problem (with either the lecture or the audience).

Verbal feedback is a more subtle way to gauge the audience–lecture connection. Based on the types of answer to the lecturer's questions and the type of questions posed by the audience, it is possible to estimate the audience's depth of understanding. In other words, it is a subtle way to poll the audience without interrupting the flow of the lecture.

Feedback as observed by the noise level in the lecture room is the most obvious sign of a lecturer–audience disconnection. Lecture noise often starts at just a few points in the room, but quickly spreads if not confronted directly. The following chapter discusses several techniques to control the audience and focus their attention towards the lecture.

6.6 Chapter checklist

- You should pay a good deal of attention to the audience during the lecture in order to fine tune and adapt your lecture appropriately

- The most important indicator during a lecture is the amount of audience noise or chatter, which can be indicative of confusion, boredom, or fatigue
- The type and number of questions asked during the lecture is another indicator – questions that are out of place or too simplistic might be indicative of a lecture–audience mismatch
- A very useful but often overlooked indicator is the look on the faces of the audience – fatigued eyes, tilted/napping heads, or excessive yawns are obvious signs that your lecture is not going well and that drastic action needs to be taken
- The attention level of the audience during the lecture, after rising slightly and reaching a maximum during the first few minutes, gradually decreases for the remainder of the lecture
- After some time (typically, between 10 to 40 minutes), the audience is no longer able to process and take in information from the lecture in a usable and productive way

7 Effective tools/tricks to energize your lecture

There are times when even the most experienced lecturers get stuck, forget their line of thought, are unable to convey an important concept, or simply may start to lose their voice and get tired. Should these situations occur too often, the lecture becomes useless and the audience's view of the lecturer becomes very negative. For this reason, it is often useful to be aware of certain tricks as well as tools for audience regeneration.

In this chapter, we will go over a subset of these tricks/tools including confronting mistakes, lecture polls and surveys, non-defensive lecturing, as well as breaks, jokes, and other fun distractions. Used effectively by a novice lecturer, these tools and tricks can make the difference between a good lecture and a disaster. However, if these tools and tricks are used effectively by an experienced lecturer, they can make the lecture an unforgettable experience for the audience.

Effectively, in this chapter we assume that all that could be done regarding the lecture quality metric has been done. However, the lecturer quality and the audience quality metrics can be energized and increased by certain methods that will be explained in the following sections.

7.1 BE SINCERE, CONFRONT YOUR MISTAKES DIRECTLY

One of the greatest mistakes that lecturers, speakers, or teachers often make is the erroneous belief that they should never make mistakes, and if they do they should cover them up. Making mistakes shows that we are human, and as long as this is handled correctly it will allow the audience to become more at ease with the lecturer. With that said, the number of mistakes should be minimized at all costs, since they interrupt the flow of the lecture. This section discusses the post mistake crisis management strategies that can be enacted to handle and dissolve lecturer errors properly.

So, our situation is as follows: you are in front of 200 students giving a lecture and you realize that something that you have talked about for the past 10 minutes contains a fundamental error. What do you do? Your first option is to ignore the whole thing, which is obviously an even bigger mistake to make (it also damages/reduces your perceived degree of competence). Your second choice is to quickly admit the error, and move on. While this is better than the first option, and is in fact what most lecturers commonly do, this again has its flaws because it presents an unfavorable image of you as a lecturer AND it will most likely confuse a significant portion of the audience.

When a mistake happens, there are two things to be worried about. The first is the incorrect information that the audience has received which must somehow be corrected without confusing them. The second is the trust that your audience has in you as a lecturer. The more you make a big deal about the mistake, the more the audience will understand but the less they will trust you. The more quickly and briefly you discuss your mistake, the more the audience will be confused and in the long run, the less they will trust you. It seems that for the audience's sake it is best to make a big deal out of the mistake, although for your lecturing image it is a lose–lose situation.

As always, there is a third option which can really end the confusion of the audience AND build on the trust that they have in you (as long as this option is not utilized too often). This third option is to make an extremely large deal about the mistake, more or less as follows:

1. explain exactly where the mistake occurred
2. show your disappointment in yourself
3. redo the part or the entire lecture that involved the mistake (depending on how large a mistake it was)

4. warn them that you are only human and may make further mistakes in the future. However, also tell them that whenever you do, you will always correct the mistake during the lecture and redo the lecture whenever large mistakes occur.

With this approach, the audience gets a second (and this time, hopefully correct) look at the lecture material and will understand the material equally as well or better than if no mistake had ever been made. Furthermore, your direct confrontation of the mistake, humility in responding to it, and effort in correcting it, will provide a net gain of trust between you and your audience.

Just be careful, since this approach only works well if used sparingly. Used too often, it will negatively affect the trust of the audience in you, and most likely increase the confusion of the audience in the process. Of course, the extent of redoing a lecture depends on many factors, including whether it is a one time lecture or part of a series of lectures (as is often the case in a university course). In the latter case, it is possible to perhaps redo an entire lecture, whereas in the former case redoing more than a 1–3 minute section would be unwise.

Case in point
For the typical 35-lecture semester courses that I teach at the University of Toronto, I would normally redo about 1–2 lectures either because of unintentional mistakes or because of the confusion of the students after class about the topics discussed during the lecture. So, if the need arises, it is advisable that at most 5% of the time of a lecture be dedicated to recovering from mistakes by redoing certain topics.

Another important point is that a mistake may not necessarily be a direct/obvious error in an equation or explanation. Lecturing mistakes also include going over a topic too quickly, or explaining a subject in a confusing fashion. These indirect mistakes are just as serious as direct/obvious ones, but can be harder to detect. Usually, the verbal, acoustic, and visual feedback of the audience can be useful indicators of the possibility of a mistake, be it direct or indirect.

7.2 MAINTAIN YOUR COMMAND

A lecture has many similarities to a military boot camp. For either a lecturer or a boot camp instructor, it is important to exert and illustrate command and control on a continuous basis. Doing so makes everyone fall into line and focus on the task at hand. A lack of control will introduce chaos which renders the entire experience useless. Of course, the extent and methods by which military instructors and lecturers exert control are drastically different. However, as observed by my old instructor Professor Sergei Dmitrevsky, there are elements of lecturing which can be successfully adapted from military teaching paradigms.

Perhaps the most important of these paradigms is the need for discipline. Contrary to popular belief, such discipline does not need to be enforced in a rude or aggressive fashion. It can be done gently, subtly, confidently, and with a smile. All that is required is for the audience to know that there are lines that should not be crossed.

Figure 7.1. A close look at an audience that is fully under the control of the lecturer and under captivation of the lecture.

Such lines include respect for their peers and respect for you as the lecturer. Once the lines are established (which usually occurs in the first 10 minutes of a lecture), the lecturer can maintain command of the audience with much greater ease than would otherwise be possible.

Case in point
At around the age of 5, when most kids would be busy playing with toys, I was instead eager to hear about the tactical strategies of war as outlined in a book by Napoleon, or about the Roman and Greek war strategies as outlined in the various books that my parents had in their possession. When playing with toys, I would line up all the toy soldiers and Lego-based spaceships, and pretend that I was giving them a pre-battle speech. I would then reenact the various war strategies that I had heard (along with a few modifications and counter-strategies of my own).

It is funny how some things in life never change. Now when I give a lecture to my students, I am in effect preparing them for battle. It is a battle with the complexities of careers and with life. On some occasions, it helps me to focus my energy by remembering that the students are in fact in this life-changing and consequential battle. Doing so always makes my voice sharper, my focus stronger, my energy higher, and results in a dramatically better and more polished lecture. It also brings back a few memories from those childhood speeches that I used to give!

7.3 DEMOCRATIC LECTURES
To confirm whether or not the audience is confused, tired, or bored, a poll is often one of the best and most truthful methods of verification that can be used. Polls, surveys, and questionnaires are the hallmarks of democracy that work equally well for an audience as they do for an entire country. The essence of democracy is that by allowing people the right to choose their destiny, even if only given a minor choice, they will have greater satisfaction and acceptance of the outcome and direction of their destiny.

This essential idea works equally well for lectures where, by interacting with the audience and giving them a choice about their destiny, they often become more engaged and interested in the lecture. Of course, just like anything else, too much interaction diffuses the focus of the lecture.

Democratic processes in a lecture can take many different forms, including quick polls and surveys. Polls, done quickly to get feedback from the audience, are invariably utilized by lecturers at some point. However, their typical utilization is often in the realm of "have you seen this topic?" While such topical polls are a good idea in general, they do not exploit the maximum potential of a poll to energize the audience and provide useful information for the lecturer. Questions such as "how many people are bored with this topic?", or "how many people are confused about this?" can be used on occasion to 1. show the audience that you care whether they are bored or confused and 2. to obtain somewhat reliable real-time information about the state of the audience and the effectiveness of your lecture.

Quick polls work equally well for one-time lectures and speeches as for multi-lecture courses. In multi-lecture courses, however, short written surveys can also be used as a type of "super" poll. This survey, which should be anonymous, will give you a much more reliable status reading on the lectures than a non-anonymous poll during the lecture. Furthermore, surveys can convey important lecturing tips for your specific audience who may be less or more advanced, interested, and engaged than you might think.

Course evaluations are one such type of survey, except that these evaluations which occur at the end are entirely useless for the course itself. Mid-term course evaluations are now common at many universities. However, more common course surveys (once a month) are also a viable option as long as the surveys are short and to the point (i.e. just a series of 5–6 multiple choice questions).

It actually helps if these course surveys are interesting and funny, instead of dry and to the point. The audience will have an easier time reading and filling out the survey if they get a sense that they are enjoying the process.

Case in point

For all my courses, I conduct a monthly written anonymous survey (held near the end of the lecture, at which time I leave the lecture hall to let the students write the survey without any intimidation). This short survey consists of 6 multiple choice questions and a comment box. I have found the information obtained from these surveys profoundly useful in modifying and fine-tuning my lectures. My survey questions and format appear below:

Professor Aarabi's monthly teaching assessment survey

Please answer the following six multiple choice questions. Put your answer at the indicated location on the back of this survey. Also, please provide additional comments on the back of this page. **Do NOT put your name on this survey.** All surveys will be completely confidential.

1. In comparison with other courses that you have taken or are currently taking, how would you rate the difficulty of the subject matter in this course?
   A) This course is a joke! I am completely bored!
   B) It is pretty easy. I am sometimes bored.
   C) The difficulty is the same as other classes. I am comfortable with the difficulty level of the class.
   D) This course is rather difficult. There are certain concepts that I do not understand.
   E) HELP!!! I have no idea what is going on in class!
2. In comparison with other courses that you have taken or are currently taking, how would you rate the pace (i.e. the speed of the lectures) of the class?
   A) Too slow. This is not kindergarten; he can go a bit faster.
   B) Rather slow. I would like to see a slightly faster pace.

C) I am quite comfortable with the current pace of the class.

D) The pace is a bit too fast for me. There are certain concepts that
   I do not understand because of the fast pace.

E) Too fast! One second we are on chapter 1, the next second we are on
   chapter 3. Professor Aarabi goes through chapters as fast as he goes
   through Nesteas!

3. Does Professor Aarabi offer enough office hours? More generally,
   compared to other courses, how much effort does Professor Aarabi
   put into the class?

   A) Not very much. He could certainly put in more effort or offer more
      office hours.

   B) He tries, but his effort level could certainly improve.

   C) I don't know enough to answer this question, but I believe he puts in
      an adequate amount of effort.

   D) He puts a bit more effort than other Professors.

   E) He puts a lot more effort than other Professors. I am very pleased
      about the amount of time that he has set aside for his office hours.

4. How do you find the clarity of the lectures?

   A) The lectures are not clear at all.

   B) The lectures are somewhat clear, but they could be improved.

   C) The lectures are okay.

   D) The lectures are very good.

   E) The clarity of the lectures is excellent.

5. Are there enough examples and applications discussed in class?

   A) No, there are no examples/applications at all.

   B) There are a few examples here and there, but a lot more is needed.

   C) The number of examples and applications is good, but it could be
      slightly improved.

   D) The number of examples and applications is perfect.

   E) There are too many examples and applications.

6. How would you rate the overall teaching ability of Professor Aarabi?

   A) He can't teach!

   B) He is okay, but needs to work on his teaching.

   C) His teaching ability is average, the same as other Professors.

D) Professor Aarabi is a very good teacher.

E) Professor Aarabi is an excellent teacher.

Answers: 1.__ 2.__ 3.__ 4.__ 5.__ 6.__

Name one thing that you would like to know about Professor Aarabi:

Comments :

A final note about surveys is that unlike polls, where the results are clearly evident to the audience, survey results may not be as evident. As a result, it is very important always to share the results of each survey with the audience while using the results as a basis for changes to the delivery of the course/lecture. An interesting and very beneficial side-effect of such surveys is that they increase the confidence and trust of the audience in your lecturing ability. For example, if half the audience mention on the survey that the lecture is going too slowly, and the other half say that it is going too fast, then without a survey the entire audience would be disappointed with the pace of your lecture. After seeing the survey results, they may still be disappointed, but they would then understand why you are progressing at a moderate pace.

7.4 BE ON THE OFFENSIVE

A lecture can in many ways be considered to be a game or contest, with two opposing sides: the audience and the lecturer. The lecturer's goal is to convey ideas and information and to teach the audience, while the audience subconsciously fights against the intake of this information and ideas. In fact, any excuse including boredom, lack of confidence in the lecturer, or lack of interest in the subject matter can be used (again, not on purpose, but subconsciously) by the audience to focus elsewhere and not on the lecturer.

As a result, just like a game, contest, or competitive sport, it is the lecturer's job to be on the offensive and to engage the audience. Certainly, one definition of being on the offensive is to be prepared and to have an exciting lecture with a good deal of audience interaction. However, the meaning of "being on the offensive" in this section really

Figure 7.2. A zoomed out view of a lecturer who is engaging the audience, resulting in a non-defensive and effective lecture.

should be taken in the literal sense. The lecturer should occasionally walk up to the students and up into the aisles if possible. The lecturer must try to deliver the lecture not to the audience, but to individuals; to look them in the eye and individually explain to them the new concepts and material. In other words, not to fear the audience but instead put fear in them that they may be selected to answer a question if they are focused elsewhere, or picked on if they speak during the lecture. This "being on the offensive" lecture style is usually the hallmark of the best lecturers, and to really understand its effectiveness consider the opposite, the "defensive" lecture.

A defensive lecture consists of a lecturer who does not engage the audience and who always stands behind the podium, speaking with a monotone voice, while deeply engaged in his/her own slides/boards. Such a lecturer has little awareness of the state of the audience or of the random and unrelated topics that the members of the audience are thinking about. All a defensive lecturer has is a small amount of trust by the audience in his/her expertise, and with even the slightest fumble in the words or a mistake, that trust too will be gone. A defensive lecture, like a defensive war, or like a defensive football/hockey/basketball game, can only end in disaster. Victory requires one to be on the offensive. In our case, making the audience pay attention to the lecture (and not elsewhere) requires the lecturer to engage the audience, to talk to them, to walk close to them, to be aware of their state of mind, to look them in the eye, and to constantly put within them the fear of being chosen or spoken to. This fear arises only for those in the audience who would normally not pay attention, and will force them to focus on the lecture just in case they are selected.

Case in point
My first stint as a lecturer was for EE260, an undergraduate/graduate course on signal processing at Stanford University. What was even more challenging than the 100 or so students in the class was the fact that the lectures were taped and televised live on the Stanford Instructional Television Network (SITN) AND online. I had the terrifying option of watching myself at night in my dorm room on television with sufficient clarity to notice my hands and voice tremble because of my nervousness. What made the whole ordeal even worse was that I had literally begged several professors at Stanford to let me teach this course, partly to get teaching experience, but primarily to impress a girl (who could also watch me on SITN)! During the dark times of those first few lectures, I would best describe myself as extremely defensive. I was always at the blackboard and would not usually step more than a meter away from it. I also preferred to look at the board instead of the students.

As the quarter went on, and as it became clear that things with that girl I was trying to impress were not going to work out, I slowly realized that things could not get any worse. As a result, I started to become more confident, and my lecture style became less and less defensive and moved more onto the offensive. Having SITN feedback for the lectures allowed me to fine tune individual lecturing points. I also noticed a gradual shift away from the blackboard and closer to the audience. By the end of the Winter 2001 EE260, I was engaging the audience, asking them questions, and was far more confident than I could have imagined. Perhaps the best lesson that I learned that winter was the transition from delivering defensive lectures to lectures that are more engaging and on the offensive.

7.5 BREAKS, JOKES, AND OTHER FUN DISTRACTIONS

Perhaps the most important tool at the disposal of a lecturer is the ability to distract the audience by either questioning them about non-content-related issues (such as what some in the audience might be talking or laughing about), making jokes that are either unrelated or indirectly related to the lecture content, or, by taking a 2–10 minute break to allow the audience to relax, recover, and refocus on the remainder of the lecture.

Regarding these actions, the ability to make funny, relevant, and correctly timed jokes is a skill that a lecturer may or may not possess. We will proceed on the assumption that the lecturer may not have this tool at her/his disposal. We will now focus on methods of distracting the audience from the lecture content. The first question to address, of course, is why? Why would we ever want to distract the audience from the lecture?

The reason behind this relates to the audience quality metric that was defined in The Equation. The audience quality, which is directly related to the attention span of the audience, is in fact a time-varying metric. As shown previously, the attention span of the audience generally decreases with time during the lecture, and after some time T_{max}, the lecture is no longer very useful.

Distractions, be they unorthodox questions, breaks, or jokes, take a certain amount of time away from the lecture. However, they also refocus the energy of the audience by erasing in part their boredom and lack of focus on the primary content of the lecture. This is very similar to a body builder who, after a heavy bench press set, takes a break to relax the muscles and stretch before the next set. This

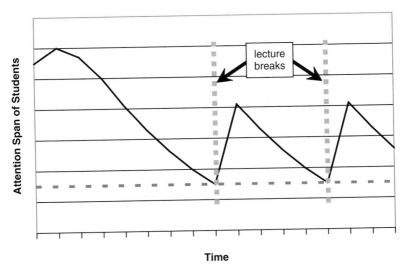

Figure 7.3. This graph illustrates the attention span of a typical audience as a result of short breaks during the lecture. Notice that the area under the attention span graph is significantly higher than if no breaks were taken, even if the breaks take away from the overall length of the lecture.

stretching and relaxation is essential to the successful outcome of the body builder's efforts. In a very similar way, by resetting and refocusing the audience, it is possible to get more out of them in a way that is illustrated graphically in the figure above:

By having occasional distractions, the focus and attention span of the audience resets. Now, the overall information that is transferred to the audience for the entire duration of the lecture (i.e. the area under the graph) is much larger than would be the case without any distractions, *even* if these distractions take some time away from the lecture.

The issue that is most relevant at this point is the value for T_{max}. In most practical lecturing situations, T_{max} will depend on the type of audience, their level of interest and match to the lecture topic, as well as their general level of fatigue.

Case in point
For my courses at the University of Toronto, I have found experimentally that T_{max} has a rough average value of around 20–25 minutes. In

other words, during a 50-minute lecture, there is a need to have a single break in the middle of the lecture (at around 20–25 minutes into the lecture). For 110-minute lectures, the students get two 5 to 10 minute breaks.

I should also mention that the first time that I tell the students about the break, I always explain the rationale behind it, including illustrating the graphs that have been shown in this section. This explanation always makes an interesting initial lecture and is an effective way of breaking the ice between myself and the students.

Assuming that a suitable value for T_{max} has been determined, the next issue concerns the style of the distraction. Questioning the audience for the sake of changing the focus of the lecture can be an effective method of distraction. Most typically, questions posed to those in the audience who speak during the lecture kill two birds with one stone by both indirectly warning the audience members who talk often and changing the focus of the lecture in a subtle but often unnoticeable way. Questions directed to those who are laughing, or sharing notes, or even doodling can be equally successful if tastefully done. Care must be taken to ensure that the questions are asked in such a way that they do not humiliate the audience or make them feel uncomfortable (which is not in any way, shape, or form, acceptable). This especially applies to those in the audience who might be sleeping, the questioning of whom must be done with the utmost gentleness and care.

Perhaps a simpler and safer solution is the break. A break is simply a 5–10 minute pause in the lecture during which the audience can talk to their friends, go to the washroom, sleep, or perform any other action of their choice. This is also a great time for a lecturer to rest his/her voice, clear the boards, etc. A break can be very successful since no comic talent is needed, as in the case of jokes, and no finesse is needed as in the case of questioning the audience. The only thing required for breaks is to explain the reasoning behind them at the beginning of the lecture (or, for multi-lecture courses, at one of the earlier lectures). Once the audience understands the philosophy of the break, they will most often view it and react to it in a very positive way.

7.6 PERSONALIZATION

An effective way of establishing strong lecturer–audience ties and at the same time having the unique ability to change gear and reset the mood of a lecture is by personalizing the lecture. One way of doing this is by learning the names of a few audience members who tend to be more social and outspoken. By doing so, they can essentially be your partners in crime, by allowing you to interact with them at will during the lecture. This also has a side benefit that since these individuals tend to be more outspoken audience members, by knowing their names and interacting with them during a lecture you in effect force them to be nice.

This personalization makes the lecture fun, unpredictable, and exciting. And although it is only practical with a few audience members, it will send a signal to the rest of the audience that they may be next. This instills a slight fear of disturbing the lecture, and a feeling of excitement that they are a part of the show (i.e. the lecture) as well.

Case in point

For the Engineering Science courses that I teach, I always get a student photo sheet with the names and pictures of the students before the start of the semester. This allows me to study the list prior to the start of classes. During the first few lectures, I seek to identify the more outspoken, loud, and social students and remember their faces. In fact, as soon as I find such students, I quickly glance through the photo sheet in the middle of the lecture and circle their picture indicating that they have been marked for my personalization experiment!

In subsequent lectures, after I have memorized the main details of the social students, the information is put to good use. While this utilization could simply involve looking at the students and calling out their names, my personal preference is to rely on more exotic techniques. For example, for maximum dramatic effect, I may notice an identified student talking while I am writing on the board. In such a case, I simply write on a part of the board "John Doe, please refrain from disturbing the class" or something similar. It takes the audience a few seconds to understand what is happening, but they are always amazed by it because they had not realized that I knew who in the class was talking let alone their specific name. Another approach is to stop the lecture, walk up to the specific student, and ask: "John, is there something you want to share with the rest of us?"

I am sure that there are many other fun ways of utilizing such student information. However, care should be taken in making a positive student match. Just imagine how bad it would look if a student was misidentified!

7.7 CHAPTER SUMMARY

In this chapter, we learned about the methods of lecture revitalization that are both subtle and effective. The important points of this section included confronting your mistakes directly during the lecture, by:

1. explaining exactly where the mistake occurred
2. not being afraid of showing disappointment in yourself
3. redoing part of or the entire lecture that included the mistake (depending on how large a mistake it was)
4. warning the audience that you are only human and may consequently make further mistakes in the future. However, also tell them that when you do, you will always correct the mistake during the lecture and redo the entire lecture whenever larger mistakes occur.

Other important points in this section included conducting quick polls to get a sense of audience interest and the effectiveness of the lecture, as well as larger audience surveys for more long term feedback. These surveys are essential for any lecturer to know what actions are best suited for a given audience. Also it is important to realize that as a lecturer you must continuously take action to make sure that the audience understands your lecture (this is another way of saying "don't be on the defensive"). These actions can include personalizing the lecture and interacting with the audience or telling jokes to momentarily distract the audience. Another useful action is to give your audience a short 5–10 minute break every 20–25 minutes in order for them to catch their breath and regain some of their energy/focus. This is especially important if your lecture is highly theoretical and difficult to understand, and less important for more common and general purpose lectures where the audience will likely be paying more attention.

7.8 Chapter checklist

- Whether you are faced with audience fatigue, boredom, or a general lack of interest in your lecture, there are actions that you can take to turn the situation around
- The first and most important thing is to ensure that you have the audience's trust – honesty and the ability to admit and confront your mistakes go a long way in securing this trust
- As a lecturer you must keep your command and control of the audience and the lecture – the moment that you lose control, the lecture is no longer useful
- Feedback from the audience regarding your lecture (from sources such as surveys and polls) can be a very useful method of detecting possible flaws and maintaining control over the audience
- Another method of maintaining control consists of shocks such as breaks, jokes, or other fun distractions
- Breaks after every 20–25 minutes of lecturing are very useful since they partially reset the fatigue level of the audience and increase their attention span
- Jokes and other fun distractions can be an effective way of both humanizing yourself and personalizing the lecture

8 Common mistakes that turn good lectures into disasters

Now that we have discussed in detail what you should do in a lecture, it is equally important to remind the reader of some bad habits that can often make a lecture a disaster. These significant errors in lecturing decisions are actually quite common. There is a tendency in all of us to either pack too much into the lecture or to go too quickly through a topic which the audience does not understand. In the following sections, we will look at these common mistakes, and come up with solutions as well as a list of DON'Ts.

8.1 OVERCONFIDENCE DISGUISED AS ARROGANCE

Confidence is a great thing, especially when it comes to giving lectures. The line between proper confidence and overconfidence, however, can sometimes be blurred. It is essential to keep in mind that overconfidence is often seen as arrogance, resulting in a strained audience–lecturer connection and lack of belief and interest in the lecture. This is especially the case, I have noticed, for successful (or not so successful) lecturers who are at the beginning or towards the middle of their careers.

It is useful to note that lecturing is in part similar to being a politician. The audience must be engaged and satisfied with the lecture or else they will not give you their attention, just as they would not give their vote to a politician. As a result, just like a politician who needs to work on getting the support of the people, so should a lecturer for her/his audience. Overconfidence and arrogance can be the downfall of both politicians and lecturers. Examples concerning politicians are plentiful; just one look at a nightly newscast will show this. Examples concerning lecturers, while less newsworthy, also tend to occur far too often.

The root cause of this overconfidence is the erroneous belief that the individual has attained some form of perfection. A belief that the lecture will be a success no matter what the lecturer does. This, in turn, reduces the lecturer's feeling of responsibility, the need for rehearsal and organization, and attentiveness to the audience and their needs. Even if the lecturer is still able to give acceptable lectures, the audience will detect a hint of arrogance which can make them combative and distracted.

Good lecturers need a sense of humility and humbleness just as much as they need focus, rehearsal, and all the rest of the important elements mentioned before. As lecturers, we need to realize that the process of lecturing is just that of a more experienced human talking about her/his experiences with other humans. Respect of the audience by the lecturer is just as important as respect of the lecturer by the audience.

Case in point

Early on in my career, after receiving a few teaching awards followed by several consecutive years with good teaching evaluations, I started to arrogantly and incorrectly believe that I could not do anything wrong when it came to lecturing. My attitude during the lectures that year changed, starting from the way and manner in which I would walk into the class, all the way to how I would answer the students' questions.

This attitude quickly resulted in some tension during the lectures. While I soon learned about the negative aspects of my actions and tried to correct them, their destructive consequences remained throughout the term. The end result was less effective lectures, a strained student–lecturer relationship, and reduced course evaluations. This was a mistake that I learned never to repeat again.

8.2 PACKING TOO MUCH INTO THE LECTURE

The most common lecturing error made by many lecturers (often professors in classes) is to put too much information into the relatively narrow channel between the lecturer and the audience. It must be understood that the lecturer is an expert (usually) at the top of the field that she or he is lecturing about. The audience is often not as much of an expert as the lecturer, hence they require more time and more examples to digest the information that is being presented to them.

Is it better for the audience to understand 70% of the lecture perfectly, rather than to see 100% of the lecture but not really understand any of it? Clearly, close to perfect understanding of a subset of a lecture is almost always desirable, and as such this suggests that, as much as possible, lecturers need to be aware of how much information their audience can take in. In other words, as a lecturer, pick a few of the most important points and focus on these. It is unnecessary and impossible to cover all the nuances of a single topic. This focusing is essential for intelligible, useful, and effective lectures.

Case in point
I have found that in typical undergraduate university courses, such as probability and applications, digital logic, or computer organization, it is possible to reduce the amount of information by 50% while keeping the content more or less intact. By focusing on the fundamentals of the course, and avoiding some of the confusing side issues (not all, but only some), it becomes possible to teach a much more coherent, focused, and thought provoking course. When the students are bombarded by facts and equations, they lose their ability to understand anything (even the fundamental concepts) and the whole lecture becomes useless.

By the way, just as an interesting note, what I have found is that lectures become overreaching not by trying to cover too much material, but often by trying to cover too many nuances of a single topic. This is especially true in the first few lectures where, by virtue of covering the first few chapters of the textbook, they tend to have more fat and less direct/important/relevant content. This is a great time to speed up by covering only the bare minimum, saving time for the more important sections of the course.

8.3 MISCONNECTION WITH THE AUDIENCE

There are two types of audience misconnections that can occur during a lecture. The first is an audience–lecture misconnection, which is the inability of the lecturer to prepare a lecture that is well suited to the audience's level of expertise. A second type of misconnection, which can be called an audience–lecturer misconnection, is when the lecturer is unable or unwilling to build a personal connection with the audience. We will now explore each of these types of misconnection in more detail.

Imagine delivering a graduate level theoretical physics lecture, in all its mathematical glory, to a class of grade 1 elementary school students. Clearly, there is a mismatch between the audience and the lecture. If the lecture was significantly simplified, or if the audience age/grade was increased, then the level of mismatch would decrease. However, how can we identify the point at which the mismatch disappears? The correct answer in most lecturing situations is one of precedence: if a topic has been lectured to an audience of the same age/group in the past, then it is assumed that lecturing to the same age/group would be appropriate. This commonly used definition, however, may be incorrect in situations with either a changing audience demographic or past errors in incorrectly presuming the suitability of a topic for an audience.

While precedent gives us a rough idea of the type of lecture that should be delivered to different audiences, no amount of pre-planning can ensure a perfect match between the lecture difficulty/complexity and the understanding level of the audience. As such, it becomes extremely important for the lecturer to keep track of the audience during the lecture and to make dynamic adjustments as needed.

This brings us to the audience–lecturer misconnections. Good lecturers are always engaging, interactive, and have personalities and presentation abilities that resonate and are remembered by the audience. This does not mean that a lecturer must be overtly nice (although being nice certainly does not hurt), but that the lecturer must be aware and tuned into the general feelings, emotions, and needs of the audience. For example, if a lecturer receives too many simple questions or even blank faces, she/he can slow the pace of the lecture to better suit the audience. Bending and adjusting the lecture for the audience should, obviously, only go so far so as not to disturb the overall point of the lecture. However, small adjustments here and there that are based on a solid audience–lecturer connection can have a significant impact on the success of the lecture.

Case in point

While most of the Case in points revolve around my own experience as a lecturer, this example involves my observations as a member of the audience. During my undergraduate years in Engineering Science at the University of Toronto, the two most memorable lecturers that I had the pleasure of knowing were Professor Sergei Dmitrevsky and Professor Tarek Abdelrahman. Both of these astounding instructors have received their share of teaching awards and praise from the students. This is in large part due to the personal connection that they established in their lectures, but interestingly, they did so with completely opposite connection strategies.

Tarek's successful strategy involved delivering extremely clear lectures with a dash of humor that was both well timed and well placed. He was very approachable and generally established a warm and friendly connection with the students in the class. The humor combined with his clarity and sincerity made learning in his classes enjoyable and fun.

Sergei's strategy, one which he explained to me many years after when we were both on the faculty of the University of Toronto, was one modeled around a military boot camp. Having himself experienced a tough no-nonsense type of education and military training, Sergei's lectures were memorable first for the boot-camp type of atmosphere which involved some students being tossed out and many others being spoken down to, and second for the extreme clarity of his presentations. Sergei's extreme no-nonsense approach (which in some instances was so extreme that it was almost humorous, by Sergei's own intention), which scared quite a few of us undergraduate students, worked wonders for focusing our attention onto the lecture and making us study harder than we thought possible. While his connection with the audience was quite different from Tarek's, both types of lecturing presented a memorable experience for the students.

8.4 MISUSING OVERHEAD SLIDES AND POWERPOINT PRESENTATIONS

Overhead or PowerPoint slides are dangerous tools, as is all technology related to lecturing. Used effectively, such pre-generated slides can provide an innovative stimulus for the audience. However, if used ineffectively, pre-generated slides become boring, confusing, and in extreme cases depressing.

The danger, which is one that many lecturers are unaware of, is the fact that a pre-generated slide will almost always reduce the

Figure 8.1. A picture of a PowerPoint presentation that Parham gave at the Federal Parliament of Canada.

"lecturer quality" metric defined in The Equation. When a lecturer is at the board, putting energy into explaining the subject matter, pushing the chalk to draw a straight line, and waving her/his hands to graphically explain the concepts, this output of energy is registered by the audience as an increase in the capability and quality of the lecturer. PowerPoint slides and overhead slides reduce the lecturer quality since the audience does not directly see the effort put into making the slide. Even if they did, the audience would almost always underestimate the effort required to produce a PowerPoint or overhead presentation.

On the other hand, one may argue that there are cases where PowerPoint and overhead slides are used effectively. If there is always a reduction in the lecturer quality metric, then how can the level of understanding of the audience increase in certain situations?

The answer here relates to the fact that although the lecturer quality metric might be reduced (perhaps only slightly), the quality of the lecture itself, which can now include audio, images, and video, in *some* cases goes up with PowerPoint and even overhead presentations. Furthermore, by having video or audio clips, the audience is allowed a virtual break which increases their attention span. As a result, the net effect of an effective pre-generated presentation can be positive, as long as it is done correctly.

So, what makes an effective slide? The following is a set of general guidelines that will help in generating effective PowerPoint or overhead presentations:

1. Know that the lecturer quality metric is at a disadvantage for pre-generated presentations. Hence, the lecturer must visibly try to show her/his effort in explaining the points on a slide. It also helps if the points are displayed one at a time rather than all at once on the same slide.

2. Pre-generated slides should be as simple as possible with a few important points (maximum four) displayed using very large font.

Miniature fonts and/or heavily packed slides quickly reduce the lecture quality to zero. Remember to avoid writing full phrases and putting all the information on the slide. While a slide should be more or less self-explanatory, it should not replace the role of the lecturer or the lecture for that matter. Hence, always leave something to be said during the lecture beyond what is on each slide.

3. While animations are useful, they can be extremely annoying if used too often. One way of producing a visually pleasing presentation in PowerPoint is to have very subtle animations (fading, smooth transitions/movements between slides) throughout the presentation but to avoid drastic animations (blinking, constantly growing or shrinking, etc.) completely. Note that the best animations are ones which the audience does not even notice because they have become used to the subtle movements and animations commonly seen in movies and television productions.

4. Images and audio are always good (as long as they are related to the lecture).

Case in point

I will now illustrate examples of some good, average, and very bad slides. The topic will be from a lecture that I have presented at several locations (Stanford University, University of Waterloo, York University, University of California at Berkeley, as well as the Federal Parliament of Canada). For the most part, the audience has consisted of either experts or semi-experts in the field, thereby allowing for a more advanced lecture.

First, let us look at the good set of slides. The slides below set up the basic ideology for sound localization, give applications, show an interesting picture, and define some of the basic concepts in two slides. Please note that all talking points are minimal, and that the lecturer will have to go into more detail than what is on the slides (hence increasing the lecturer quality metric value).

Basic Sound Localization:

Microphone arrays can localize sound using time of arrival and intensity differences.

Applications:

 – smart rooms
 – automatic telecon.
 – robust speech rec.
 – robotics
 – other applications

Figure 8.2. A good slide introducing a topic in a very basic and fundamental way.

Enhanced Sound Localization

- N microphones, each observes a signal $m_i(t)$:

$$m_i(t) = \psi_i(\mathbf{x}, \theta) \cdot s(t - \tau_i(\mathbf{x})) + n_i(t)$$

- Define SLF to be proportional to the log likelihood:

$$F(\mathbf{x}, \theta) \propto \log p(\mathbf{x}, \theta \,|\, \{m_i(t)\})$$

Figure 8.3. A good mathematical slide with minimal detail and just a few necessary fundamental equations.

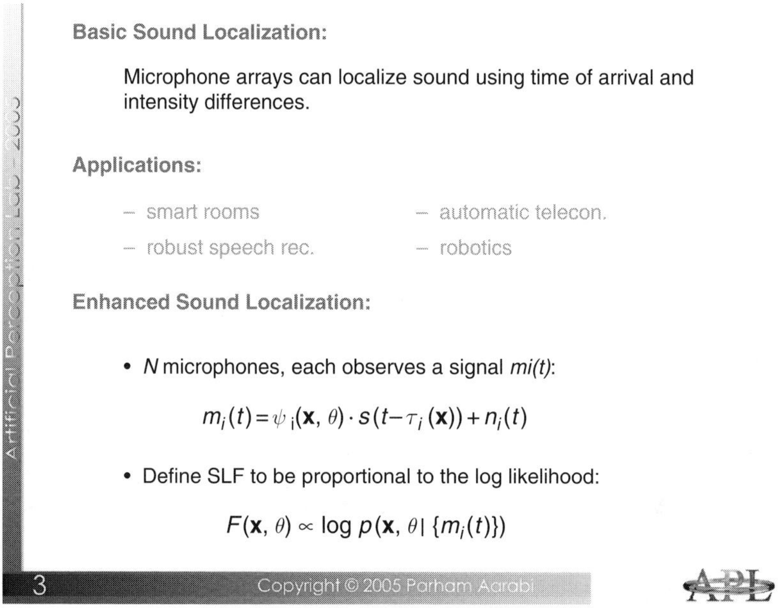

Basic Sound Localization:

Microphone arrays can localize sound using time of arrival and intensity differences.

Applications:

- smart rooms
- robust speech rec.
- automatic telecon.
- robotics

Enhanced Sound Localization:

- N microphones, each observes a signal $mi(t)$:

$$m_i(t) = \psi_i(\mathbf{x}, \theta) \cdot s(t - \tau_i(\mathbf{x})) + n_i(t)$$

- Define SLF to be proportional to the log likelihood:

$$F(\mathbf{x}, \theta) \propto \log p(\mathbf{x}, \theta \mid \{m_i(t)\})$$

3

Figure 8.4. A bad combination of the previous two slides, with no pictures, a reduced font, and more overcrowded.

Now, we will look at gradually worsening slides. First, let us look at an average-to-bad slide that contains the above points:

As can be seen in this slide, the information is too tightly packed in (although not horribly) and is of a small font size. While this slide in itself is not a disaster, it has no image, is not exciting, and does not do as good of a job as the previous two slides.

We continue to the final horrible version of the same slide. This slide is as tightly packed as possible (you would be surprised how many lecturers like slides such as this!!!!). Furthermore, every sentence and every topic is fully explained with even the minute grammatical details. This forces the lecturer to just read from the slide, and minimizes the value of the lecturer. As tempting as having such complicated, ugly, and intense slides might be, *always* stay away from this type of slide!

Basic Sound Localization:

Microphone arrays can localize sound using time of arrival and intensity differences. Time delays are utilized since if a microphone is further away from the source, then the sound will arrive at a later point.

Applications:

   – smart rooms, where sound localization can help track people
   – automatic teleconferencing, where we steer a camera to the speaker
   – robust speech recognition, where we use the speaker location to only listen in that direction
   – robotics, where a robot would follow the speaker

Enhanced Sound Localization:

• We assume that there are a total of N microphones, each observing a signal $mi(t)$. Our simple model can be illustrated as follows:

$$m_i(t) = \psi_i(\mathbf{x}, \theta) \cdot s(t - \tau_i(\mathbf{x})) + n_i(t)$$

• Now, we define the SLF to be a function proportional to the log likelihood of the probability of the speaker being at any location given the data:

$$F(\mathbf{x}, \theta) \propto \log p(\mathbf{x}, \theta \mid \{m_i(t)\})$$

3

Figure 8.5. A nightmare slide, including full sentences, extremely small font, and extensive overcrowding. While this is perhaps the most typical style of slides that lecturers use, its only value to an audience would be that of a sleep aide.

8.5 CHAPTER SUMMARY

In this chapter we have observed some of the common mistakes made during lectures. These mistakes simply involve lectures that are too heavy or difficult for a given audience, lecturers who are out-of-tune with the audience, as well as presentations that are too boring and/or monotone due to the overt usage of complex slides.

This is of course only a small sample of the many errors and mistakes that can be made which negatively impact the level of understanding of an audience. However, the most important first step in avoiding and resolving errors is the awareness of the lecturer about potential problems. Due to the dynamic nature of lectures, it is generally possible to overcome most errors and to correct the direction and the impact of the lecture. This can only be done through the focus and awareness of the lecturer. Of course, having presentations, slides, or blackboard lectures that are fun, exciting, and not excessively boring or complex does not hurt either!

8.6 Chapter checklist

- Arrogance on the part of a lecturer can severely harm the lecturer–audience relationship
- A lecturer who is perceived to be humble allows the audience to ask questions and to interact more freely and with greater appreciation of the lecturer
- Packing too much into a lecture will slowly corrode the lecturer–audience relationship
- Lectures must be tuned to the abilities of the audience – an audience–lecture mismatch can have very negative consequences
- Bad overhead or PowerPoint slides can also have a significantly negative effect
- Slides should be simple and to the point, with only the minimum necessary amount of detail
- Too much detail, especially if it is unrelated to the main issue under discussion, will bore, confuse, and scare the audience away from the lecture topic

9 At the end of the lecture

A very important and often neglected aspect of lectures is the way in which the lecture ends. Again, if we consider the television show analogy, the ending of most episodes is exciting, concluding, happy, and sometimes, involves a cliffhanger to motivate the viewers to watch further episodes. A lecture, in a very similar way, needs to end on a positive note while providing ample motivation for the audience to get excited about what they have just experienced. As a result, great care must be taken in the last few minutes of a lecture to end on a high note. For example, rushing to finish an example quickly due to the lack of time can have very negative consequences, as is the case when a topic causes confusion in the last moments of the lecture with no time to clarify the situation or to answer questions.

These issues will be explored in detail in the following sections.

9.1 DO NOT RUSH

The end of a lecture, unlike the beginning, needs to be a calm, relaxed and smooth event. It must finalize the topics of the lecture with a few moments to spare for further clarification. A very common mistake is to rush in these last moments due to an imperfectly executed lecture plan. However, what does rushing at the last moment really accomplish? Is reaching a lecture coverage goal worth the audience confusion that would result? Or, is reaching this goal more important than the negative perception of the lecture by the audience when they leave the lecture room?

The best thing to do in the last few minutes of a lecture is to slow down. If the lecture plan has not been perfectly executed, then there is nothing wrong with leaving part of the lecture for another time. Even if it is a one-time lecture, the ending of the lecture can be quickly

Figure 9.1. An image of the end of a lecture with some of the audience members starting to leave the lecture hall.

summarized in a high-level fashion without confusing the audience or rushing them at the last moment. In essence, when you are short on time, finish your lecture by switching to a high level lecturing mode instead of rushing forcefully through the remainder of your detailed lecture.

It should also be noted that during the planning stages of a lecture, you should keep in mind that a shorter than expected lecture with a possibly lengthy question period and an early send-off is far better and more likely to be appreciated by the audience than a longer than expected lecture that is rushed in the last few minutes.

Case in point

A lecture is a dynamic and unpredictable event. Even after perfect planning, the number, length, and type of questions asked during a lecture can never be estimated perfectly. This makes the exact timing of the end of a lecture unpredictable.

What I often do during the planning stages of my lectures is to have multiple ending points that occur at different stages. For example, I might plan for a core lecture that would occupy half the allotted time, and then follow that with a series of practical examples and case studies which I would only do if there was enough time. This way, near the end of the lecture, based on how slowly or quickly things have progressed, I choose an appropriate end point that does not leave an example hanging but does leave several minutes for questions.

9.2 LEAVE ENOUGH TIME FOR QUESTIONS

A usually successful lecture planning strategy is to plan for a slightly shorter than usual lecture while allowing for about 5 minutes of questions at the end. This often works extremely well, since the question period can be used to either clear up any confusion that the audience might have, or to end the lecture a few minutes early. Either way, the audience greatly appreciates the outcome.

The question period can also be used to state high level comments, historical perspectives, as well as points of experience regarding the material discussed in the lecture. These can be effective tools at both humanizing the material while at the same time personalizing the lecture.

Often, it may be necessary to extend the question period beyond the time allotted to the lecture. While this can be a tedious and lengthy extension to the lecture, it is usually a very good way for the audience to ask their own personal questions and end any confusion. This is also a good way for a lecturer to strengthen the audience–lecturer relationship.

9.3 RETAIN CONTROL UNTIL YOUR LAST BREATH

In the mayhem that often ensues near the end of a lecture, it becomes extremely difficult to maintain control over the audience. From the audience's perspective, they have sat and listened attentively for a long time. Restlessness usually follows such periods of attentiveness, resulting in an eagerness to escape from the lecture room. Such restlessness is very contagious. When a single member of the audience starts to pack up their items and look towards the door, several more quickly do the same, followed by many more. Before you know it, the

Figure 9.2. A lecture room with a restless audience that is eagerly awaiting the end of the lecture.

entire audience is engulfed in mass unrest and at that point, whether the lecturer wants it or not, the lecture is over.

The sudden onset of mass restlessness can occur without notice and as much as 5–10 minutes before the end of the lecture. When it does occur, it neutralizes the ability of the lecturer to interact effectively with the audience since most of the audience's focus turns towards the conclusion instead of the content of the lecture.

It is here that you as a lecturer must assert yourself so that you can maintain the focus and attention of the audience until the last moment of the lecture. Always remember that a good lecture must end on the final phrase of the lecturer instead of the restless choice of the audience.

Case in point
Often in the last few minutes of my lectures the audience becomes restless and tries to force an end to the lecture by starting conversations and producing noise. I react to this initially by halting the lecture for a few seconds and making sure that the audience knows that the lecture will be over when I say so, not when they do.

After a few minutes, depending on how tired or restless the audience might be, they again start conversations and begin to pack up their notebooks. At this point, I remind them of the importance of what I am talking about and the urgency regarding which it must be stated (for example, if a lecture precedes a lab which requires specific knowledge). I then ask them for just a few more minutes in order for me to finalize the delivery of the lecture. At this point, I know that I am trying to squeeze the last ounces of energy and attention from the audience, so I resort to asking them nicely for it.

Usually, this gives me the final few minutes of calm that I need prior to the conclusion of the lecture. On most occasions, there will be enough time for questions and the exact timing of the conclusion of the lecture will not be an issue. Occasionally, when I incorrectly estimate the amount of time that a lecture will take, I reach a point in the lecture when, after halting the lecture and after asking the audience for a few more minutes, I lose control of the audience without being able to finish the lecture. Essentially, based on my experience, there is a threshold of restlessness for audience members beyond which no request, no matter how scary or nice it may be, is capable of delaying the untimely conclusion of the lecture.

9.4   REMEMBER, THE LAST COAT OF PAINT IS THE ONE
THAT LASTS

If you have ever painted pottery, or oil paintings, or even a house, then you know that the most important coat of paint is the last one. This is the coat that will mostly if not entirely affect the final color. Similarly, for a lecture, the last few moments of the lecture and the final messages of the lecturer are what the audience will remember the most. It is therefore crucial that these final messages be carefully thought out and carefully preplanned. Just as a painter may ask herself/himself about what color of paint should be used last on a wall or a piece of pottery, so should a lecturer ask herself/himself about the final messages that the audience will take away.

Essentially, these final messages must be designed to maintain the interest of the audience in the lecture even after it has concluded.

Possible ways of doing this may consist of giving the audience a difficult problem whose solution would warrant a reward, or by telling the audience a captivating and memorable story in order to relate the lecture topic to their own personal lives. The latter approach in effect ties the lecture to the individual personal lives of the audience, thereby increasing the likelihood of further thought by the audience about the lecture topic.

Case in point
In the last 6 years, I have tried a variety of lecture ending phrases and tactics with varying degrees of success. The most successful strategies seem to be the ones that involve a surprise. For example, on one occasion I told the students at the end of the lecture about the cancellation of a lab, which energized them and excited them beyond my expectations.

On another occasion, at the end of a lecture I told the students about a mid-semester pizza party, which instantaneously resulted in loud applause. The mid-semester pizza party is something that I have occasionally hosted to both personalize the lectures and to be a different and unique lecturer.

Aside from surprising lecture endings, which cannot always be achieved easily, the next best strategy is to motivate and intrigue the audience by asking them to solve a difficult question at home. Here, by telling the audience about a difficult problem that they must think about, you raise the curiosity and interest in the subject matter for a certain portion of the audience. Perhaps a better approach that will affect a greater portion of the audience is one that I learned from my colleague and friend, Professor Sam Roweis. Professor Roweis's approach consisted of putting up difficult problems related to the lecture topics and offering the audience rewards for solving them. The same approach can be used successfully to end a lecture, by posing difficult problems and offering rewards (such as a grade bonus) to the first person who can come up with a solution.

Rewards and surprises are of course difficult to provide on every single lecturing occasion. As a result, on many occasions you may have to settle for a simple "have a nice day" ending to the lecture preceded by a short summarization of the main points of the lecture. While this is nothing exceptional or memorable, it is generally good enough in most cases as long as once in a while you include more eventful endings to your lectures such as rewards and surprises.

9.5 CHAPTER SUMMARY

In this chapter, proper lecture conclusion strategies were discussed. It was stressed first and foremost that good lectures must end on a high, smooth, and exciting note, and should never be rushed or quickly and carelessly presented. Rushing and lecturing heavily during the last few minutes of a lecture will negatively impact the overall lecture experience. This is in contrast to the start of the lecture where lecturing heavily with a slightly fast pace can have positive consequences.

A smooth conclusion to the lecture can be accomplished by means of a question period which is both slow in pace and productively effective. At all costs, situations where the audience has started to leave prior to the conclusion of the lecture should be avoided, since they reduce the clout and potency of the lecturer. The last few messages of a lecture are the ones that the audience will remember best. Consequently, they must be very carefully planned and rehearsed.

9.6 Chapter checklist

- The end of the lecture is a time for calm and control, not a time to rush through a poorly planned or poorly rehearsed lecture
- The lecture pace should always be slow near the end of the lecture and should end a few minutes early to leave time for questions
- The question and answer period at the end of a lecture is a lecturer's most effective tool at combating confusion
- Control of a lecture must be kept from the time the lecturer utters the first words of the lecture until his/her last words
- A lecturer who loses control of the audience and is unable to regain this control will be perceived as weak and powerless in

the eyes of the audience, making the continuation of the lecture extremely difficult
- The last few minutes of a lecture are the ones that the audience will remember best, consequently, every attempt should be made to inspire and excite the audience in a memorable way during this period

10 The art of academic lecturing

So far we have discussed the nuances of giving a lecture in any type of setting. However, in this chapter, we will focus on the issues that can arise while teaching in university, college, or any other type of setting which requires multiple lectures and possibly a set of tests and exams.

With multiple lectures, it is possible to give one bad lecture and still recover by giving several excellent lectures afterwards. In other words, there is more room for lecturing errors than the single-lecture case. As a result, teaching multi-lecture courses is an excellent way of polishing your lecturing abilities. Furthermore, having to evaluate the audience/students gives them the incentive to listen to your lectures. This gives you a natural advantage in attracting the audience to your lectures. The exact method, type, and difficulty of the evaluation (i.e. test, exam, etc.) can be used to control the learning experience of the audience by either comforting them with a relatively straightforward test or by shocking them with a difficult test. In fact, in many cases it makes sense to use a combination of straightforward and difficult questions.

The following sections will take a more detailed look at several important multi-lecture course issues.

10.1 THE FIRST LECTURE

As with the first few minutes of a lecture, the first lecture of a multi-lecture academic course is also very important. It is during this lecture that the students in the class get their permanent impression of the lecturer. If this first lecture goes well the students will have a positive view of the lecturer and of the course for the remainder of the term. If the first lecture goes poorly, then the lecturer will need to spend significant energy to overcome the poor start.

Also, just as in the first minutes of a lecture, over performing and lecturing at a fast and heavy pace during the first lecture (or even the first few lectures) puts the audience in the right mindset. This "right mindset" consists of them being slightly worried about the course, thereby giving more attention, focus, and energy to understanding the lectures than they would otherwise.

Essentially, the first lecture needs to be your flagship lecture. It must be exciting, a little scary, memorable, and even shocking. Such a lecture will set the multi-lecture course on the right path and will make your job as a lecturer easier. To do this, the following few points may be kept in mind for optimal effect.

First, do not be conservative in delivering your first lecture. Try to go at a faster than usual pace. Try to make the lecture denser than you would normally be comfortable with. Also, thoroughly rehearse your first lecture in order to ensure clarity and crispness to the greatest possible extent. And finally, remember that the most important lecture is always the first one, just as the most important moment of any lecture is during the first few minutes.

Case in point
Personally I have found the first lectures of multi-lecture courses to be the toughest ones to deliver. In fact, I have had more disastrous first lectures than successful ones. While there are many reasons for this, the most likely explanations include an initial unfamiliarity with the audience, a slight amount of nervousness, and a certain level of discomfort with the lecturing process since it is, after all, the first lecture. In general, my successful first lectures all share the characteristics of aggressive (as opposed to defensive) lecturing, a fast pace, meticulous planning and rehearsing, as well as just a confident feeling brought about as a result of enough sleep and adequate exercise prior to the lecture.

10.2 TESTS AND EXAMS – TO KILL OR NOT TO KILL
Setting the level of difficulty of tests and exams is one of the most important choices that a university teacher has to make. The

difficulty of tests and exams is akin to the change in the national prime interest rates set by the Federal Reserve, or the pushing of the brake and accelerator pedals on a car. Setting tests and exams that are very difficult, resulting in an average significantly below the expected average of the class, can be demoralizing to the students. Then again, such tests do serve as a wake up call and shock, thereby forcing some students to study harder. On the other hand, tests and exams that are too easy, while lifting the spirits of the average student, will give a false sense of comfort. So, what is a good lecturer supposed to do?

The answer to this question depends mainly on personal preference. These discussions assume that there is a final fixed average that will be reached, no matter how difficult or easy the tests and exams are. The scenarios are therefore as follows:

1. you set an easy in-term test, an easy final exam, and adjust the final grades by lowering the class average
2. you set a very difficult test and exam, and adjust the final grades by increasing the class average
3. you set a difficult test and an easy exam, with little grade adjustment at the end
4. you set an easy test and a difficult exam, with little grade adjustment at the end
5. you set an average test and an average exam, with little grade adjustment at the end

The first strategy should be avoided at all costs. It establishes a false sense of security for the students only to be crushed at the sight of their final grades. This is morally reprehensible. The second strategy is perhaps the most common. It does however demoralize the students into underestimating their competence in the subject matter. Finally, the third and fourth strategies will not be effective because of the imbalance between the in-term test and final exam, which can significantly confuse, shock, and demoralize the students. My personal favorite method, one which I have utilized for the past five years, is the fifth approach. My midterm test grades are rarely more than a few percentage points from the expected final grade and the exam average grade.

With all this said, actually getting the average you desire as the lecturer on a test or exam is another matter altogether. Setting exams and tests which will yield the desired class average is a very complex and difficult art form. It takes years of practice, and even then there could be significant underestimation or overestimation of the difficulty of a test. The best advice is to have one or more teaching assistants try the test out with the proper time limit.

10.3   TEACHING ASSISTANTS – THE GOOD, THE BAD, AND THE …

The best resource in teaching large university courses is the help of teaching assistants. Teaching assistants or TAs usually have more personal interaction with the students than the lecturer. They will also help to lessen the burden on you as a lecturer when it comes to grading papers, tests, and assignments. As such, it is extremely important to be aware of the performance of the TAs and to take action should it be warranted.

There are generally three types of TA. There are those who are extremely dedicated to the course and the students. These are the type that deserve TA-of-the-Year awards, and whose presence on the TA team is a tremendous asset. Then there are capable TAs who either do not have a great deal of time or who are generally not very interested in the course. While these TAs are not ideal, they will do their job, and as long as they do you cannot ask for more. Finally, there are some TAs who either have very arrogant personalities or are incompetent, or both. These TAs will seldom listen to your guidelines and are generally like a cancer on the other TAs. Such TAs must be directly and quickly confronted and prevented from ruining the course. As such, getting feedback from the students on the TAs on a consistent basis is essential when teaching a large course with numerous TAs.

Case in point

In the past few years, I have had the pleasure of knowing some extraordinary TAs as well as the misfortune of some below average TAs. Particularly, I have noticed during the grading of labs (which is done in the presence of the students), several TAs whose interaction and opinion of the students has been unnecessarily harsh. Accordingly, I always have a small TA training session to explain to them the point of grading students (i.e. it is not to demean or scare them but to give them a fair evaluation of their understanding of the lab). This has resulted in inflation of the lab grades (which has been easily offset by a slightly harder midterm and exam). However, it has had the positive effect of making the students care more about the content and the fundamental point of the labs than the lab grades themselves, since the labs are likely to appear in some way on the midterm and exam.

This strategy of having leniently graded labs whose content appears on the midterm and exam was suggested to me by Professor Zvonko Vranesic, a renowned University of Toronto author, researcher, and lecturer, with whom I had the pleasure of co-teaching a course several years ago. The Vranesic Lab-Lecture model will be discussed in more detail in the next chapter.

10.4 MULTI-SECTION LECTURING ISSUES

When different instructors are lecturing in parallel sections of a course, with common tests and a common exam, numerous issues can arise. These issues usually arise only when both (or all) instructors are relatively inexperienced. They can range from the simple student shifting problem (where students go to a specific lecture due to the time of the lecture or the style of the lecturer) to more serious problems including rivalry between lecturers. Other common problems that can arise include lecturers who firmly believe in different lecturing philosophies, thereby resulting in conflict from which neither side is willing to back down. If certain students ever get a sense of such rivalries and issues, just like sharks who taste blood in the water, they will make great efforts to worsen the situation.

Resolving and avoiding such problems requires constant communication between the instructors. Furthermore, it requires an open mind on all sides to listen to each other's ideas and teaching styles. Finally, a very important issue for university administrators is to ensure that at least one senior individual is among the set of instructors for every course, which is the best way to prevent problems in the first place.

With all that said, a final comment regarding multi-section lecturing issues needs to be made. While it is important, as a lecturer, to compromise and to listen to the other instructors, you must only do so when the compromise does not hurt the students. If you believe that the actions of another instructor will, either through ignorance or malice, negatively affect the experience of the students and their lives thereafter, it is your responsibility and duty to do your utmost to dissuade and divert the other instructor from his/her course of action.

10.5 ALWAYS REMEMBER THE PRIMARY GOAL

Your duty as a lecturer is to ensure that most students understand the course material and the fundamentals of a subject, and that they are graded fairly based on their level of understanding. They must know enough to be able to build on this knowledge in subsequent courses and throughout their professional careers. They must also be judged in a fair manner that neither overestimates nor underestimates their capabilities with respect to their peers. This is the primary goal, and everything else including your relationship with the TAs, with the other instructors, as well as the amount of time and effort that is required, while important, is a secondary issue.

Case in point

Many years ago, when teaching one section of a multi-section course, a conflict arose between me and another instructor regarding the final averages of our sections (we had decided to equalize all section averages at the beginning of the year). The other instructor wanted an average for all sections that was 10% lower than my suggestion. Such a lower average would result in 50 more students (out of 300) receiving grades below the "A" range. This would in turn affect their chances to attend professional and graduate schools. It should also be mentioned that the departmentally recommended, expected, and historical average for this class was exactly what I was suggesting. Hence, a significantly lower average would have been unfair to the students.

After discussing the issue with the other instructor, I realized that I had a choice. Either I could avoid a conflict and go with the lower average, or, I could keep my suggested average and go through a long and possibly bitter conflict. It is funny how morality and ethics are such theoretical concepts until you are faced with tough choices like the one I had to make. The other thing that weighed on my mind was that while both of us were relatively new instructors, the other instructor was approximately 10 years older than me. Perhaps part of the reason for ignoring my suggestions might have been due to my youth.

Now, if I did not know the students in this class by name and if I had not talked with many of them about their lives, dreams, and aspirations, then maybe, just maybe, I would have chickened out to avoid the conflict. However, on that terrible day after my discussions with the other instructor, the notion of this primary goal of always caring about the best interests of the students took hold in my brain. I realized that, while I can be selfish at times, I could not damage (even slightly) the future of those 300 students just to make my immediate future more comfortable.

Long story made short: a conflict did arise and eventually the final average that I was suggesting was adopted, but only after the department chair had intervened. I regret that I could not be more convincing regarding my discussions with the other instructor and I regret that a conflict had to arise. However, I never regret the decision that I made, to put the well-being of the students and the fairness of the final grades ahead of everything else.

10.6 POST-LECTURE INTERACTIONS

As a lecturer, you initiate and establish a mutual relationship and connection with the audience. This connection does not and should not end when the lecture ends. Lecturers should be open and friendly (obviously to a point) when it comes to helping the audience members and answering their questions.

In a similar fashion, it is very important to establish a connection with the students who have attended your multi-lecture course. For an entire academic term, these students have sat in your class and listened to your every word (if you did your job right, that is!). As a result, it is essential for you to allow them the possibility of interacting with you in the future. This could, in its simplest form, involve exchanging greetings in the school cafeteria and buildings. However, this may also include you hiring them as summer students or potentially as future graduate students. It is also a good idea for you to tell the best students in your class about the possibility of reference letters and to send them an email of encouragement after the tests and especially after the final exam.

Positive words and positive feedback, no matter if it comes during the course, after the course has just ended, or years after the students have graduated, never hurts.

Case in point

One of the hallmarks of my teaching style is to send students in my classes unique emails regarding their performance on the midterm tests and final exam. These emails range from extremely positive to constructive. I have found that this goes a long way towards establishing a strong lecturer–audience connection and to show them that I want them to succeed and that I am willing to help them if they are willing to put the effort that is required into the course.

I should also mention that while during the first few years of teaching I would send a unique email to each and every student, after my class sizes rose above 200, sending such unique emails became impossible. As a result, I now have about 10 categories of students based on their performance on a test or exam to each of which I send a unique performance-based email.

10.7 COURSE LOAD ASSESSMENT

There are a variety of methods for evaluating the teaching load of different courses. These methods range from a simple count of the number of course sections being taught to a more detailed weight based on the complexity, level, and enrolment size of the courses. It is important for lecturers to know what the typical load of a course is and to allocate sufficient hours per week for preparation and rehearsal.

Generally, for every hour of lecture approximately an hour of preparation and rehearsal time is required. The actual time will vary depending on the experience of the lecturer and her/his familiarity with the course topic. Aside from this preparation time, the enrolment of a class also determines the load on the lecturer, and contrary to the assessment methods at most universities, the relative load of a course is logarithmic with the number of enrolled students.

Case in point

Based on my experiences teaching courses with as few as 5 students and with as many as 400 students, the relative load of a course with a total of X students can be roughly modeled as the logarithm of X (i.e. $\log_{10}X$). Note that since this is a relative metric, the actual base of the logarithm does not matter and a base of 10 is chosen here for simplicity. For example, a class with 10 students will have a relative load of 1, a class of 100 students will have a relative load of 2, and a class of 1000 students will have a relative load of 3. It is important to note that this load assumes that there are sufficient TAs for all cases of X (at least 1 TA per 10–20 students) in order to handle grading and to help students during their contact hours. If this were not the case, then there would be a linear term (linear in X) in the course load model as well.

The relative course weight as described here is proportional to the number of hours that are required solely for contact, preparation, lecturing, and the administrative aspects of a course. Ideally, universities should use fairer metrics such as the one proposed here in order to evaluate the teaching load of various courses by taking into account issues such as enrolment size. Regardless of what they do, it is important for lecturers to keep this in mind when selecting what type, size, and number of courses to teach.

10.8 ACADEMIC PRESENTATIONS

Academic presentations, such as those given at conferences, or as invited guests at universities or research institutions, are similar to a typical academic lecture. In both cases, clarity of presentation and understanding of the subject are extremely important. Rehearsal and careful preparation are also needed no matter who your audience is. However, they differ in several subtle ways, including the fact that the main goal for the two situations is slightly different.

In a lecture, the goal is to convey information clearly, directly, and without any confusion. In an academic presentation, the goal is also to convey information but to do so while impressing the audience. The need to impress the audience will require using techniques such as the Widrow Maneuver, or showing impressive pictures and videos in order to illustrate your competence in a given subject matter. This brings us to another difference between academic lectures delivered to students and academic presentations given to researchers.

For an academic lecture given to students, the default assumption among the audience is that you are competent until proven otherwise. For an academic presentation to senior researchers at other universities, research labs, or a conference, the default assumption tends to be that you are incompetent until proven otherwise. Even for well known and senior researchers, I have observed far more rigorous and heated discussions following their presentations than in a typical academic lecture delivered to students. For junior researchers, it is almost a guarantee that they will face at least a few tough questions and some criticism from the audience during an academic presentation. This brings us to the following unforgettable, humorous, and yet unfortunate case in point.

Case in point

The first conference presentation of my career was in Atlanta, Georgia, in June 2000. Since I had very little lecturing or presenting experience prior to this date, I was understandably very nervous. Nevertheless, the idea that I was presenting, which consisted of using distributed sensor arrays to process sound and attenuate noise, was a good idea which later became the basis of my doctoral dissertation. However, my nervousness would eventually get me into trouble on that day.

My talk was at 8:30 AM EST. Having flown from California the prior evening, I was accustomed to the Pacific Standard Time at Stanford. Hence my talk was effectively at 5:30 AM according to my internal body clock. Knowing this, I tried to sleep earlier than usual the night before, only to result in sleeplessness for the entire night. To this day I remember when I first looked at the lecture room that I was supposed to present in just a few minutes before my talk. It was a large lecture room filled with over 300 experts in acoustics and signal processing, almost all of whom were older than me (I was only 23 years old at the time). Everything would have been fine had my nervousness combined with a stupid action not ruined the day.

In order to overcome my tiredness due to lack of sleep, and my nervousness, I drank a large cup of black coffee. Since my talk was about to begin, and since this initial cup had no effect, I had a second large cup of black coffee. With still no effect, I had a third and fourth cup, and then I was called up to the podium.

I went up and started presenting. The presentation was going very well until about the 5 minute point. At that moment, my hand began to shake and my eye started twitching. As I was starting to experience caffeine poisoning up on the podium during my first academic presentation, I kept going forward with the talk. To this date, I wonder what the audience thought as they saw my voice getting more stressed, my movements about the podium becoming more erratic, and my speech getting faster and faster. By the time the talk had ended, the full effect of the four coffee cups had hit me. During the question and answer period following the talk, I am not sure of the things that I said but I do remember the look on people's faces as I was answering them. Following this "eventful" presentation, I was unable to sleep for two days because of the caffeine overdose. Since that day, I have tried to stay away from coffee almost entirely, especially prior to any lecture or presentation!

10.9 CHAPTER SUMMARY

This chapter focused on issues related to lecturing multi-lecture courses. The first issue discussed was the effect of tests and exams on the audience. Easy tests and exams give hope to the students but also make some students lazy, while difficult tests and exams shock the students, some into losing hope, and others into studying. Although the difficulty of these tests and exams should be set by the instructor based on the type and state of the audience, it was suggested that an average test and exam which both match the desired final average would be best.

The next issue was that of teaching assistants who are a great asset to teaching and who have an extremely important role to play in the course. However, care should be taken to ensure that the good TAs are properly acknowledged while the bad ones are steered in the right direction. When dealing with TAs or even instructors of other course sections, it is important to keep in mind the well-being of the students above all else.

10.10 Chapter checklist

- Your first lecture is the most important one, so every attempt (from careful planning to extensive rehearsal) must be made to make this an exceptional lecture
- Tests and exams of a multi-lecture course must be designed in a way to both challenge the students in the class while at the same time not destroying their interest and hope regarding the course

- TAs are a lecturer's most important ally but they should be told explicitly what is expected of them and what their exact role in the course will be
- Teaching multi-section courses can result in numerous unwarranted struggles and arguments, but at all times, the well-being of the students should be the primary goal of all lecturers
- An effective way of personalizing an academic lecture is by interacting freely with the students after the lectures and during office hours
- For planning purposes, the overall load of a university course can be modeled approximately as being logarithmically proportional to the number of students in the class

Making lectures come to life through labs

No matter how effective the lecturer or the lecture, nothing can replace the experience and knowledge gained when an audience member obtains first-hand knowledge through practical experiments and labs. This obviously applies to multi-lecture courses, but can even involve simple in-class experiments conducted during the lecture. No matter how and where these labs/experiments are conducted, they are a unique and essential tool in the teaching process. In this chapter we will take a closer look at how labs should be successfully organized, calibrated to the lecture, and evaluated.

11.1   THE POINT OF LABS AND PRACTICAL EXPERIENCE

The clarity and focus that has been repeated throughout this book for lecturers is equally important for labs. The point of labs is to take a somewhat complex concept or idea, simplify it in a way that can be touched and felt, and to allow the students to understand the basis for the idea through hands-on experience.

Occasionally, understanding comes from initial success. However, more often, true and deep understanding comes only after repeated failures. You can be told many times that an electronic device may act a certain way, and you can certainly learn a few things from setting up that circuit in the proper way. However, it is all the circuit forms that are non-functional and all the ways in which common mistakes could be made that will truly have an impact on what you will remember and what you understand.

In life, whether you want to learn to become a better lecturer or learn about electronic circuits, you need to try, experiment, and play with your own medium of focus. The point of labs and other forms of practical experience is to enable students to perform these experiments

Figure 11.1. A view of a typical lab at the University of Toronto.

Figure 11.2. A view of students engaged in individual computer laboratory experiments.

and to, in essence, play. As a result, the most successful labs tend to be ones which allow creativity and do not penalize non-conformity.

11.2 THE RELATION BETWEEN LECTURES AND LABS

It is important to note that lectures and labs should ideally complement one another. One does not really replace the need for the other, though the two do need to be tightly interconnected. The most ideal experience for students is to hear an explanation of the fundamental concepts in the lecture followed by hands-on experiments in the lab.

It is often a mistake to have labs whose focus is unrelated to the lectures. Some lecturers attempt to teach more by covering some topics in labs and other topics in the lectures. This "pumping as much information as possible" notion is almost always wrong since it destroys the essential point of labs. The labs should give students the practical experience needed to better visualize and understand what the lectures discuss. In other words, the presence of labs does not really allow more material to be learned, but for the material covered in the lectures to be learned better.

Case in point

Sometimes, it is possible to directly incorporate labs into lectures. Perhaps the best example that I have seen has been that of Professor Brendan Frey. Professor Frey, aside from being a world-renowned researcher and a great friend, has a very unique and successful teaching methodology. In teaching a course on probability and statistics, he often includes small probability experiments directly into the lectures. His experiments are chosen for their surprise factor, including counter intuitive probability experiments such as the Monty Hall problem or the probability of matching audience birthdays, whose high probability (for medium to large class sizes) is almost always a surprise for the audience until they see the simple mathematics behind the surprise.

These group labs have two very positive effects. For one thing, they make the lectures more fun and add yet another tool to shock and motivate the students. Also, they provide a unique experimental perspective to look at the lecture material, thereby inspiring and exciting the students.

11.3   WHAT A LAB SHOULD NOT BE

The worst possible mistake that can be made regarding labs is to emphasize grades over understanding the fundamental concepts behind the lab. Labs are meant to be a playground of sorts, allowing students to express their curiosity in their own way. While clear guidelines and explanations of the lab are always required, rigid constraints which prevent students from learning, or rigid grading methodologies which require and enforce a specific report form over substance, often destroy the whole point of labs.

It is imperative that the importance of the labs and their purpose be clearly conveyed to the TAs who supervise and grade the labs. Occasionally, TAs who lack experience or clear guidelines tend to increase the rigidity and the harshness of the evaluation schemes of the labs, partly because it inflates their stature during the moment of evaluation. This, of course, is not true for many TAs, but for some, it is unfortunately the reality. As such, you as an instructor need to be on top of this situation by attending labs often, talking consistently to the TAs, getting feedback from the students, and taking appropriate and quick action when required.

Case in point

When I was a first-year undergraduate student, the labs for the first-year physics course often seemed to be a grade-lowering tool rather than a useful complement to the lectures. I would often lose marks on my lab reports simply because the form that I had chosen would not match what the TAs had wanted. However, what the TAs had wanted did not in turn match with what the instructor had told us. This great chain of miscommunication between the instructor and the TAs continued over many other courses and several years. These experiences have collectively resulted in my current view of labs and lab-lecture methodologies.

In recent years, for the undergraduate courses that I teach, I always have an initial discussion with the TAs regarding the goal of the labs. I then repeat these goals in the presence of both the TAs and the students. I require the TAs to have a mentality of taking grades away ONLY when there is cause to do so, rather than giving grades only when certain rigid milestones have been reached. I also tell the students that the form, format, and style of their report will not in any way affect their lab grades as long as their reports are concise and clear. I even suggest that they use the back of an envelope, if they wish, to summarize the results of their experiments! To this day no one has taken me up on my offer, though they do get my point.

Now, there are some instructors who would be shocked by the above paragraph. These instructors have an infatuation with complex forms and lengthy reports. To them, it is just as important, if not more, for the students to learn good report writing skills rather than the fundamental ideas and concepts behind the labs. While I respect their opinion, I truly believe that their mentality is ABSOLUTELY wrong! Report writing IS important, but not to the point of consuming the technical goal of the labs.

11.4   THE VRANESIC LAB-LECTURE MODEL

The evaluation of labs, on the one hand, forces students to focus, prepare for, and understand the lab while at the same time biasing them to work towards better grades instead of a deeper understanding of the lab concepts. Labs that are graded too leniently will inevitably be regarded as a joke, while labs that are too tough will depress and deter the students' curiosity.

An interesting solution to this problem was proposed to me by my colleague and mentor Professor Zvonko Vranesic. This award winning instructor and author of several textbooks on digital logic and computer systems, with whom I had the pleasure of co-teaching a course early on in my career, suggested that our course labs be graded very leniently but for their contents to appear on the course tests. In fact, the point that the term tests would include a significant portion of questions directly from the labs was very clearly and repeatedly conveyed to the students.

The term test that year (and in all subsequent years) took the form of a combination midterm and lab-test. However, elements from the labs eventually found their way onto the final examination as well. In my past experience with lab oriented courses, this "Vranesic" lab-lecture model (where the labs are for understanding important concepts whose ideas and elements will be directly tested on) has worked extremely well. Some students do suffer since they do not take the labs seriously. However, most students do take the labs very seriously especially since they are told by more senior students to do so! As a result, they focus on all the important concepts of the labs without worrying about the specific grade of a single lab. This outcome is the closest to an ideal lab-lecture synergy that I have ever experienced.

11.5 LARGE SCALE LABS

Today, at almost all universities in the world, labs consist of simple and repetitive tasks that allow students to learn the fundamental concepts of a given subject. There has however been some push in recent years towards more coordinated and large scale labs that share elements across multiple courses. There is room for optimism regarding such large scale labs, since the energy and problem solving skills that would be required as a result of the coordination would teach the students far more than just simple self-contained labs.

The trick here is to preserve the basic teaching elements of labs while allowing the hard work of the students to be directed towards a greater project. While for each course and subject the ideal solution may be subject-specific, there are a few facts that the reader may find of interest in the following case in point.

Case in point
In the past five years, over 1000 University of Toronto students have either taken lab-based courses that I have taught or have done their undergraduate design projects and theses under my supervision.

During this time, I have always wondered about the amazing projects that could have been completed if the time and energy of these students was focused towards a single significant goal. It is interesting to note that in the past few years I have tried to do this with my own research group. Instead of several graduate students each pursuing their own unique and unrelated projects, most of my more recent graduate and undergraduate research students are now focusing on the different parts of a single tough and significant research project.

In order to illustrate the power and potential of coordinated and focused labs in a more clear fashion, consider the case of Electrical and Computer Engineering students at the University of Toronto. Approximately 1200 students are enrolled in the four year electrical and computer engineering program. On average each student spends at least 10 hours per week on labs and design projects. During the course of the academic year, this works out to a total of approximately 300 000 man-hours. Now, a typical engineer performing work at the same technical difficulty as these labs and design projects would make at least \$30 per hour. With this very rough calculation, every year the potential financial value of the time and energy of these students would equal almost \$10 million dollars, enough to pay for the tuition of each and every one of the students!

11.6 CHAPTER SUMMARY

The fundamental point of this chapter was that labs can be an effective teaching companion to lectures as long as they are conducted, supervised, and evaluated correctly. The best labs are those that resemble a playground of sorts rather than a rigid exam. As a result, the guidelines and grading evaluations for both the supervising TAs and the students need to allow for unexpected lab ventures as well as nonconformist lab reports. It must be made clear that it is the ideas behind the labs that matter instead of the number of pages in the report or the title and ordering of the sections. One way to do so, which was coined as the Vranesic Lab-Lecture Model, is to grade the labs very leniently while testing heavily on their concepts in the course tests and exam.

11.7 Chapter checklist

- Labs are an important component of academic learning since they provide hands-on experience about the subject matter
- The one and only goal of labs is practical experience, not grade deflation
- Ideally, for maximum effect, labs should be interwoven with the material and discussions of the lectures
- Labs should be graded on the amount of information understood by the students and not by the beauty or look of their lab reports
- Learning to write good lab reports can come easily if the lab is fully understood, whereas, learning the lab can rarely come from writing a tightly structured lab report
- A successful method of lab evaluation, called the Vranesic Lab-Lecture model, involves leniently grading all labs and placing lab related questions on course tests and exams
- While difficult to do, large scale labs, which focus the energy of the students towards a large scale and significant goal, can have tremendously positive consequences on the mindset of the students

I2 Lecturing in non-academic contexts

Classroom lecturing is a relatively easy task because there are usually several chances to get things right and perhaps more importantly because the audience is very much dependent on the lecturer. Non-academic lectures, especially business lectures and talks, tend to be bimodal. Some are filled with useless acronyms and facts, without any point or focus. Others have such a high threshold for success that even seasoned academic lecturers would find success elusive. In this chapter, we will take a closer look at some of the key issues and common mistakes that are made when presenting in business and professional settings.

12.1 THE BUSINESS PRESENTATION

Business presentations share many elements with a typical academic lecture, including the need for clarity, audience understanding, as well as maintaining control over the presentation. Business presentations, however, because of the dynamic and broad nature of the audience, require a more thorough background and have to be more to the point than a regular academic lecture. Furthermore, due to the fact that business presentations are often given to colleagues or bosses (versus lecturing to students who, even if they hate the lecture, must understand the contents in order to prepare for the exam), there are pressures present which would normally not exist in the academic world.

With that said, the essential ingredients of a good presentation are almost identical for both business and academic lectures. In both cases, preparation and rehearsal are extremely important. The nervousness that arises as a result of presenting to your boss is similar to the anxiety produced by lecturing to 200 angry students. And, the

Figure 12.1. A view of a close-proximity business presentation, requiring quick lecturing reflexes and very brief yet informative points.

need to remain plain spoken and focused are just as important in the business world as in the academic world, if not more so.

Feedback regarding a lecture or presentation can also be just as effective when delivering a single 30 minute presentation as compared to delivering 30 one-hour lectures. In the former case, however, the feedback needs to be acquired much more quickly (within minutes) and acted on almost instantaneously. Such feedback can usually be obtained through visual and acoustic means as described previously or by the nature of the audience questions. Given enough practice this can be done easily. However, given the time limitations in obtaining feedback and acting upon it, the level of alertness required is a little bit higher than that necessary for academic lectures.

In fact, if we were to summarize this section, we could simply say that presenting in a business setting is just a more stressed and

Figure 12.2. While business presentations and academic lectures share many common traits, presenting effectively to a business or even professional audience requires a higher level of alertness and a more audience-centric lecturing approach.

compressed version of giving an academic lecture. Some of the key differences that arise as a result of this compression will be discussed in the following sections.

12.2 THE PERFECT ANSWER

A very important issue for both academic and non-academic lectures is the way in which audience questions are answered. In most academic settings, questions arise as a result of student confusion. These questions are generally (though not always) friendly, mild, and non-aggressive. In other academic settings (such as during a conference presentation) or in professional/business settings, questions are sometimes asked only to test your confidence in your presentation. When answering these aggressive questions, great care needs to be taken to solidify your hold on the audience. Surprisingly, the best answers for all types of questions, aggressive or not, always have the same general form.

The first step in answering a question is to understand the question itself. It often helps to repeat the question for the audience using a simple choice of words to make sure that 1. your audience understands the question, and 2. that the individual asking the question has no objection to your interpretation of what was asked. The entire process of repeating the question and answering it should be done in a slow, eloquent, and carefully spoken manner to illustrate the focus that you are dedicating to the question. In other words, you are zooming verbally and focally onto the question.

The next step is often believed to be a long and detailed answer to the question. However, a long and detailed answer is almost always a mistake and a source of further confusion. The best answers are often very short and to the point. If the answer to a posed question is yes, just say yes, and do not waste any time, mental focus, and energy. If the audience wants further clarification or details, they will ask. Providing such details automatically is never required and often not advisable.

The final step in answering questions is to ensure that the answer addresses what the questioner has asked. This illustrates in part that you care about their question, and also shows your control over the audience and the ensuing discussions. Sometimes, it is also useful to predict future questions that may arise as a result of current or past questions. In such cases, it may be a good idea to ask and to answer a question yourself. This is especially useful regarding controversial issues or the weak points of your presentation, since by asking the question you can lessen the negativity of the topic while at the same time telling the audience that you are not running away from it.

12.3 THE ACRONYM SHIELD

One of the best signs of the intellectual vulnerability of an individual is their reliance on big words, acronyms, and catch phrases to argue their points. In business settings, it is not uncommon to hear sentences like "our AMT is low hence the DWS system must be employed with a QRS to increase the SPQR yield ..." Obviously, there is a line beyond which the over utilization of acronyms makes everything nonsense. In my experience, certain individuals use acronyms as a shield to mask their lack of understanding and ignorance.

The points stressed in this book about honesty and clarity are just as important in business/professional settings as in academic settings. When giving a presentation, talk, or even during a regular conversation, it is extremely important to maintain your clarity of speech for the benefit of your audience. Confusing them might make you feel good for a short time, but such a feeling of fake superiority fades quickly. It is much better to explain everything to them clearly and argue your point effectively, resulting in a true feeling of success. Also, if you are unsure about something, instead of becoming defensive and throwing out acronyms and catch phrases, do not feel afraid to say that you are unsure or that you do not know. The ability to admit when you are uncertain shows a much higher level of confidence than an acronym shield designed to mask your uncertainty.

Case in point
During the past few years, I have had the pleasure of serving on the board of advisors of several start-up companies. While the degree of useless information in the board meetings has in a few instances been high, there was one meeting that was in a class all by itself. During this meeting, an individual was making a presentation to the board as well as to a few guests. His presentation, which should have been a simple 2-minute fact-based presentation, lasted over thirty minutes filled with acronyms and useless information. Actually, the only reason that the presentation did end after half an hour was at the insistence of several board members including myself.

It is interesting to explore what was going on in the brain of this highly educated presenter. His presentation was based on utter nonsense, yet he was very calmly explaining and defending his points. It was as if he was trained to explain nonsense in a very professional fashion, disguising it with acronyms. This is a point that recruiters in companies need to be very careful about. The danger with hiring such individuals is that once someone questions rigorously their points and virtual facts, the massive void in their brain starts presenting itself!

One thing that is interesting to mention is that the concept of an acronym shield is even utilized by people in academia. In the past few years, it is surprising how many individuals I have met, from students to professors, who rely on complex wordings to either hide their lack of understanding or to obtain a feeling of superiority. What these people do not realize is that in most cases clarity and honesty would have done them far more justice than their confusing techno-babble.

12.4 FIFTEEN MINUTES

There are some business presentations that have a significantly higher threshold for success than a typical presentation. Examples include making a presentation to your boss or company executives, or, making a company pitch to venture capitalists. In such situations, a short no-nonsense presentation is required which cuts through any nervousness or fear. Obviously, to do so successfully is a lot easier said than done.

Case in point
In early 2006, in the hopes of securing funding for a new start-up company founded by myself and one of my students, I presented our technical idea to numerous venture capital firms and investors. While their reactions ranged from those that committed funding directly to others who did not even reply to our further emails, the experience was very beneficial and interesting.

The most important of these presentations, one which was eventually unfruitful, was in front of two partners from one of the top venture capital firms in the world, Kleiner Perkins Caufield and Byers (often referred to as just Kleiner Perkins). Our 15-minute presentation followed by a question and answer session started in the right direction but at about the half way point steered in a negative direction. The main reason for this wrong turn was entirely my fault.

In reply to a question posed by Bill Joy (one of the partners of Kleiner Perkins and a founder of Sun Microsystems), I said "yes" instead of saying "I don't know." That error initiated a series of questions and unconfident answers which in the end resulted in Kleiner Perkins not funding us.

This goes to show the importance of being honest with yourself and your audience. When under pressure, it is always easier (at least initially) to fake an answer by trying to portray an all-knowing image. However, such a stance will quickly lead to a deterioration of your image as well as a loss of trust by the audience. The better and more professional answer is to step back and say that you do not know. Perhaps such a response would also anger your boss or not impress the venture capitalists, but at the very least it does not push you into a corner that will ultimately make you look far less intelligent than you really are!

12.5 PROFESSIONAL LECTURING

Aside from business lectures and presentations, there is a variety of other professional lecturing situations that may arise. One example is in the legal domain where lawyers present a case in front of a jury or a judge. Presenting to a jury is very similar to a lecture since the jury is the audience of the lawyer presenting her/his case. Again, many of the preparation and attention attracting strategies outlined in this book may prove useful in this situation.

Another profession requiring the need for polished public speaking and lecturing abilities is that of the politician. Politicians who need to make campaign speeches or address supporters in different venues are in fact lecturers. The ultimate goal of their lectures is of course to make the public believe in them and eventually vote for them, but the initial goal is to make the public understand and share their point of view. Politicians who are honest, who seem to care, who have properly prepared and rehearsed, and who have a proper understanding of what they are talking about tend to give the most successful political speeches. These are the exact same criteria for success as those of lecturers. In fact, this may be one reason why the transition from lecturing to politics or from politics to lecturing is often a somewhat smooth transition.

There are of course many other situations where good lecturing techniques would be useful. From company executives who speak in front of their shareholders to parents and family who speak at a wedding ceremony, learning to speak effectively in public AND to explain and talk in a clear and concise fashion is always very useful and beneficial.

12.6 POLITICAL SPEECHES VERSUS ACADEMIC LECTURES VERSUS BUSINESS PRESENTATIONS

As has been mentioned several times throughout this book, political speeches, academic lectures, and business presentations have a great deal in common. In all cases, there is a speaker/lecturer trying to convey a set of points, a medium (i.e. just plain speech or speech accompanied by a board, a presentation, etc.), and several audience members who try to listen to the speech with varying degrees of interest. All three activities have a need for clarity, focus, and control. The subtle differences between them, however, are worthy of a more detailed look.

The goal of a lecturer, especially in an academic context, is to make the audience understand complicated and difficult concepts. Here, understanding is the ultimate goal and the personal image of the lecturer is of secondary (although nonzero) importance. For politicians, who must captivate audiences in order to receive more votes, their personal image is of primary importance. Of course, the unfortunate reality is that many politicians place no significance on making the audience understand and place almost 100% significance on facts and lies that maximize their short term personal image. That aside, politicians and lecturers have essentially divergent goals.

Business-oriented and corporate presenters, on the other hand, are required to play a tough balancing act. They must very clearly teach their audience about a certain issue, topic, or a necessary action. Furthermore, in most situations, they must improve their own personal image and stature since business/corporate presentations can have possible career expanding or career ending consequences. In the end, the goal of a business presenter becomes two-fold: the first is a need for conveying the main message clearly and the second is the need to increase their own stature and image by means of their presentation. In essence, a business presenter is in part an academic lecturer, and also in part, a politician.

12.7 CHAPTER SUMMARY

This chapter illustrated some of the differences and similarities between academic lectures (as described throughout this book) and business presentations. In essence, the two are really the same task except that with the business presentation case there is a shorter single speech whereas for academic lectures there are multiple speeches providing more opportunity for success. In some ways, an academic multi-lecture course is an ideal and safe way to learn the give and take of public speaking as well as business/professional presentations.

This chapter also focused on the methodology of answering questions (in any setting, business, academic, or otherwise) which involves 1. repeating the question for the audience, 2. providing the most concise and correct answer possible, and 3. ensuring that the answer matches what was asked. In order to vary the pace of the lecture it is helpful to slow down on points 1 and 3, though point 2 always needs to be very quick and to the point.

12.8 Chapter checklist

- Business presentations consist of a more dynamic and difficult setting in comparison with an academic lecture – nevertheless, the points regarding clarity and simplicity are just as important here as with the academic lecture
- The response to audience questions, in any setting, should be direct, honest, and to the point
- If you do not know an answer, say that you do not know – if a direct question was asked, answer it directly – avoid going into

tangential and somewhat unrelated issues if not specifically asked

- There are some who have a tendency to use big words and acronyms in response to simple and direct questions – this acronym shield is almost always a sign of incompetence and never a sign of wisdom or knowledge
- The timing requirements of business presentations are far more constrained than in an academic lecture – usually, you have at most 5–10 minutes to make your case and turn nonbelievers into believers
- Political speeches, business presentations, and academic lectures share elements such as the need to focus and the need for clarity
- Political speeches, however, are person-centric whereas lectures are learning-centric – business presentations fall somewhere in the middle

13 The mechanics of professional presentations

There is a variety of guides, books, and theories related to giving effective presentations in professional settings. The recommendations regarding what constitutes an effective presentation are often similar. We will call presentations developed using these traditional recommendations as "classic" presentations.

Another approach that can be very effective, especially in today's age where the audiences are used to television show style emotional roller coasters and cliffhangers, is the "shock" style of presentation. These presentations have very little in common with classic presentations and are generally far more informal and to the point.

Finally, a presentation style that can generally work well in most settings is the hybrid presentation. The hybrid presentation is really a shock-style presentation disguised as a classic presentation. In essence, it maintains the effectiveness of the shock presentation while at the same time having a small amount of the professionalism and formality of classic presentations.

In the following sections, we will take a closer look at all three presentation methodologies in detail.

13.1 THE CLASSIC MODEL

The most generic and common style of presentations is that which we will refer to as classic. Classic presentations consist of well tested and well developed methods for conveying ideas during a presentation. There is a formal structure and an expected sequence of topics that need to be discussed. In the case in point below, we will examine the elements of the classic presentation in more detail.

Case in point

It should be noted that the presentation slides for the cases in point of this chapter have been reduced to only text in order to illustrate the difference between the various presentation styles more effectively. The following example was inspired by the recent Master's thesis presentation of my former student Sarah Ali. Her thesis presentation, which earned her a grade of A +, consisted of a method of representing multiple long audio segments (such as lecture recordings or songs) using short compressed audio thumbnails that were overlapped. By utilizing the human ability to recognize multiple songs at the same time, Sarah's perceptual audio compression enabled human listeners to search through a set of audio segments in half the time that would normally be required. While the slides below are not the presentation slides that Sarah used, her final presentation was most similar to the "hybrid" presentation style that will be discussed in a later section.

In this section, we will focus on a classic presentation of Sarah's thesis. Classic presentations can be defined by certain well defined

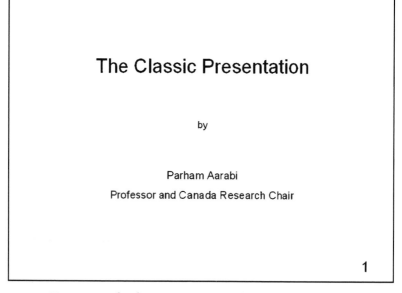

Figure 13.1. The classic presentation – Slide 1 – Title page.

features. The first is a title page that outlines the title of the presentation, the name of the presenter, as well as any affiliations, as shown in Figure 13.1.

Once the presenter has introduced herself or himself, the next typical slide is the presentation outline. Personally, there is no single slide that I find more bothersome than this one! It is entirely unnecessary for good presentations, and only becomes necessary for long and boring presentations where the audience needs to count down the sections while eagerly awaiting the end of the presentation. For reference, Figure 13.2 illustrates a typical presentation outline slide.

Once the outline of the presentation has been discussed, the presenter then moves to introduce and to motivate the audience with regard to the main points. This often involves some background material and motivational topics, and is generally the "why" of the presentation. Figure 13.3 illustrates a typical introduction and motivation slide.

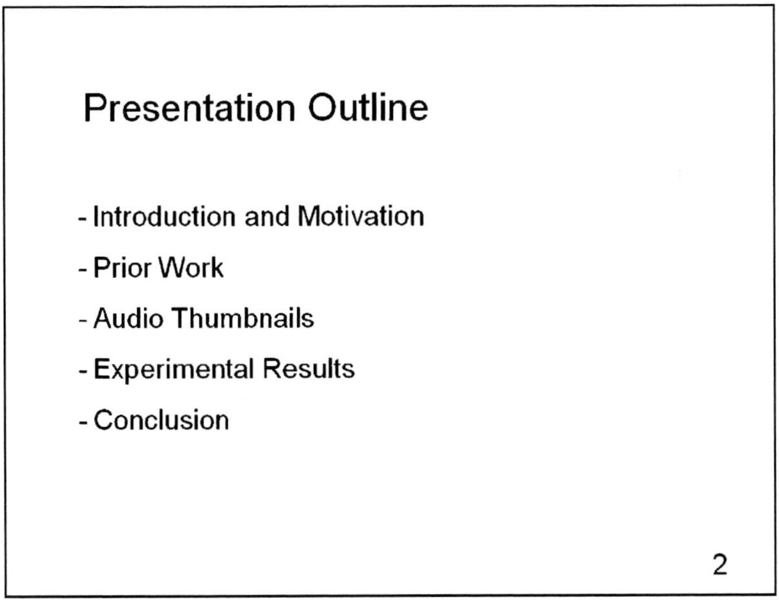

Figure 13.2. The classic presentation – Slide 2 – Presentation outline.

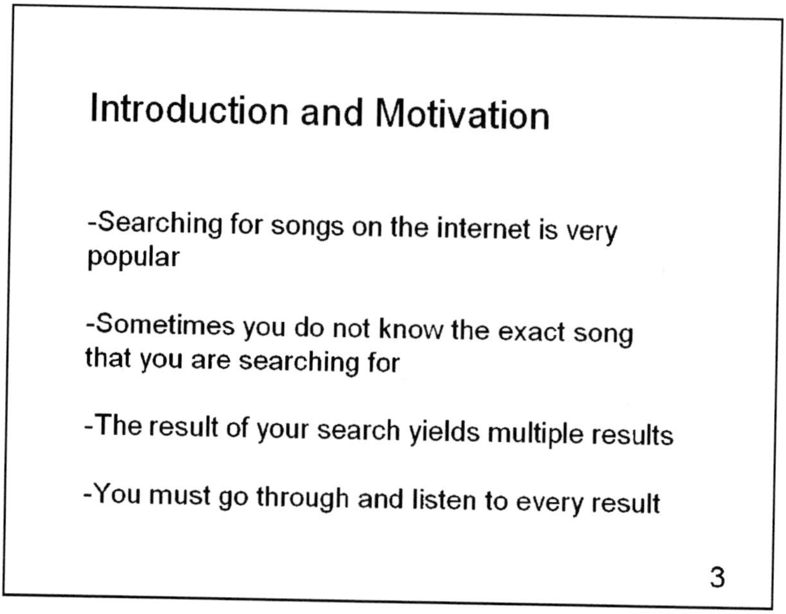

Figure 13.3. The classic presentation – Slide 3 – Introduction and motivation.

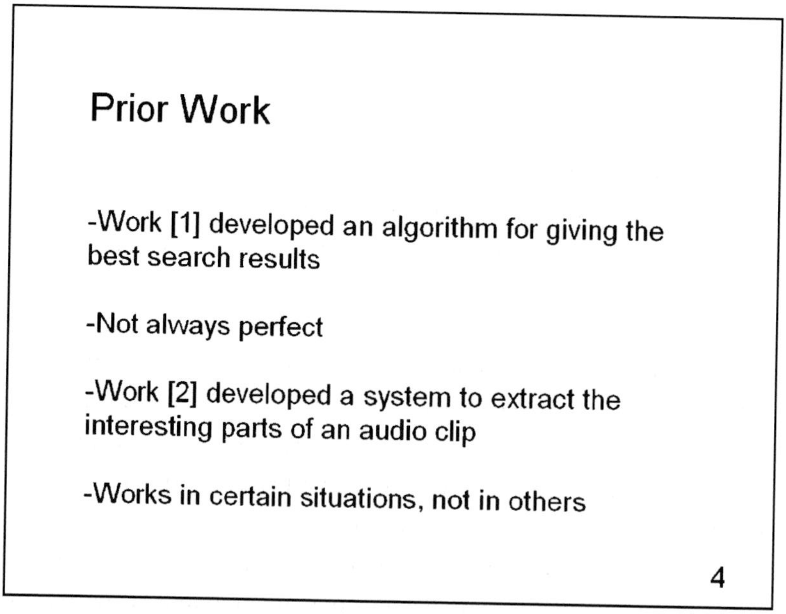

Figure 13.4. The classic presentation – Slide 4 – Prior work.

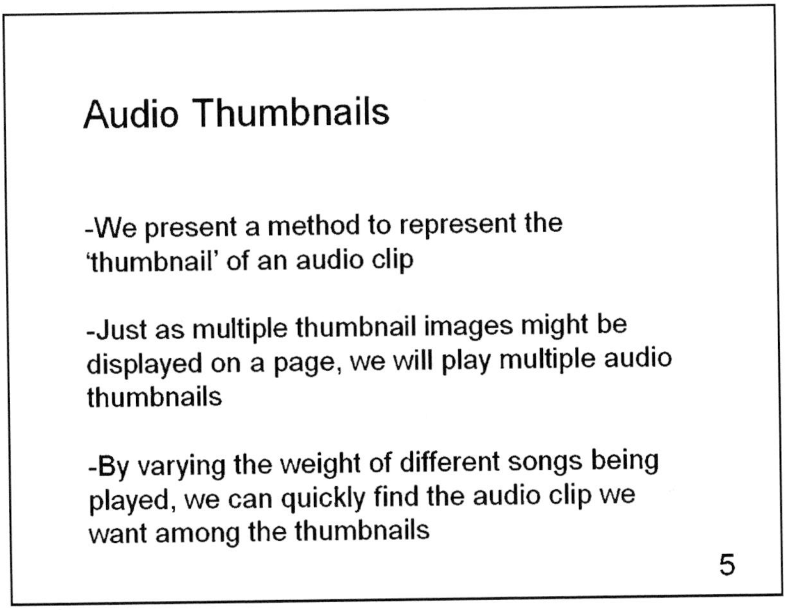

Figure 13.5. The classic presentation – Slide 5 – Discussion of the main ideas.

After the introduction slide there is usually a prior work and historical perspective slide followed by a discussion of the main ideas of the presentation. Notice that in the classic presentation the discussion of the main points occurs towards the middle and sometimes even close to the end of the presentation. Figures 13.4, 13.5, and 13.6 illustrate the prior work, main idea, and the results slides in the context of our example.

In the classic presentation, there is almost always a concluding slide that summarizes the main point of the presentation and the main conclusions. An example of such a slide is shown in Figure 13.7.

The classic presentation can be somewhat effective in many presenting situations. However, it does have limitations caused by too many constraints. In essence, it forces presenters to conform to a uniform methodology, which while being great for novice presenters who would benefit from any form of organization and structure, for most experienced presenters it can be limiting. In the next section, we

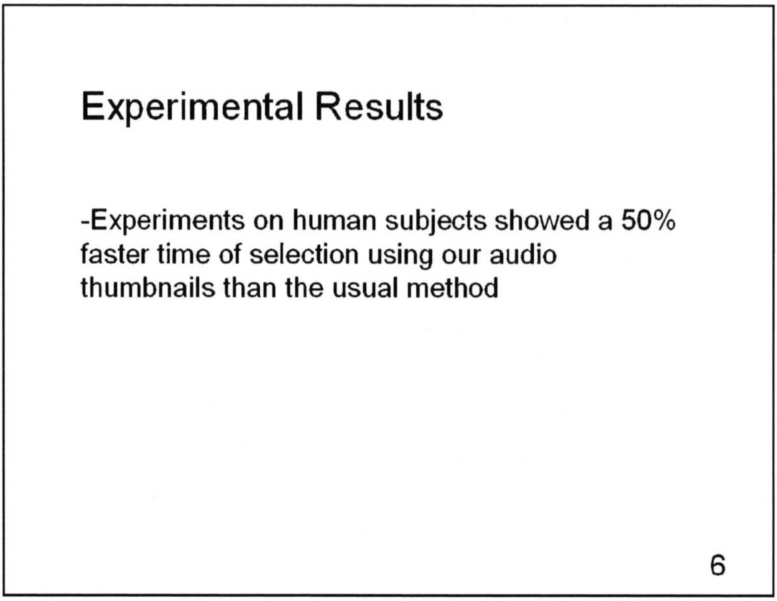

Experimental Results

-Experiments on human subjects showed a 50%
faster time of selection using our audio
thumbnails than the usual method

6

Figure 13.6. The classic presentation – Slide 6 – Results.

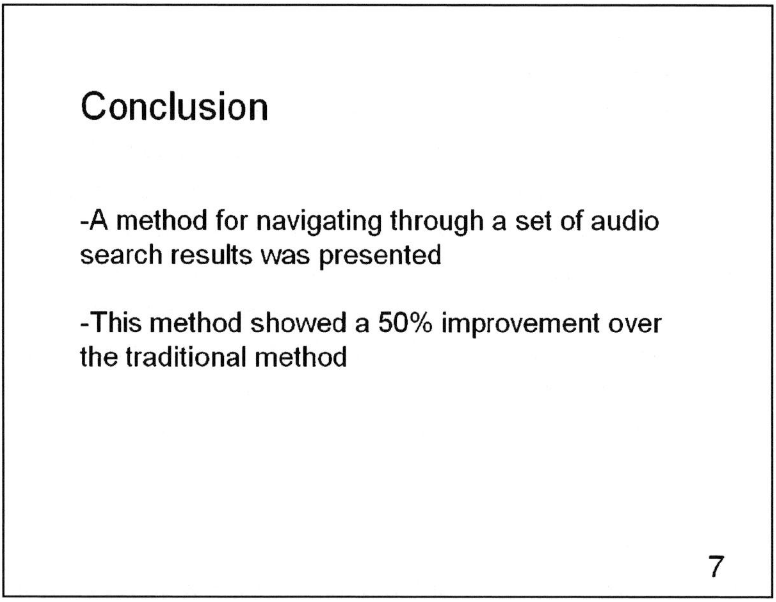

Conclusion

-A method for navigating through a set of audio
search results was presented

-This method showed a 50% improvement over
the traditional method

7

Figure 13.7. The classic presentation – Slide 7 – Conclusion.

will look at a radically different presenting philosophy called the "shock" presentation.

13.2 THE SHOCK MODEL

A presentation should not affect how we speak, but rather, how we speak should affect how we present. The classic presentation model suffers in that it places rigid rules and constraints on how a presenter talks to others about a topic. Talking, very much like presenting, comes naturally and easily unless it is constrained by unnatural rules and regulations. The obvious alternative to the classic presentation is a presentation whose purpose is solely to motivate, teach, and convey information to an audience. We will refer to this unregulated type of presentation, with no real rules or guidelines, as the "shock" style of presenting. The word "shock" is used in this context because most successful unorthodox presentations have some type of a shock somewhere in the presentation. These shocks can include a dramatic question, interesting/funny images, videos, and audio segments, or a surprising result. Sometimes, even a casual, informal, and to-the-point tone of a presentation is enough to shock the audience into paying more attention.

Since the shock presentation does not have any rigid structure, it is hard to describe it in a general sense. Some of the world's most memorable and historic presentations do fall into this category. These presentations and speeches come from the heart, are plain spoken and to the point, and often have a minimalist attitude. The goal here is to motivate and slightly shock the audience, inspire them, and to then teach them about a particular issue or topic. The following case in point looks at an example of a "shock" style presentation in the context of the audio searching example that was first used in the previous section.

Case in point

Our shock version of the audio searching presentation starts with a motivating question, as shown in Figure 13.8. This is in contrast to the

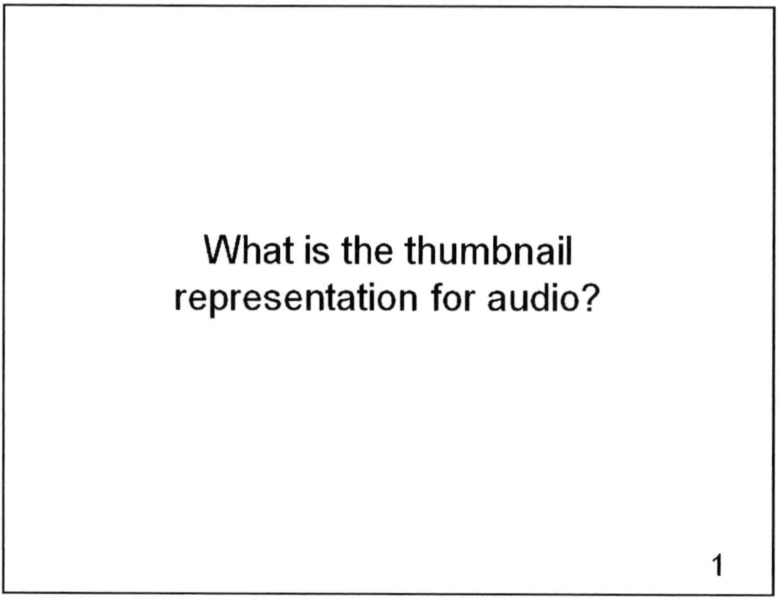

Figure 13.8. The shock presentation – Slide 1 – Opening question.

classic presentation which required several slides prior to getting to the main motivation. The general idea here with the shock presentation is to get right to the point as early on in the presentation as possible. What better way to do this than to start with the main question that has motivated the entire reason for and the entire goal of the presentation.

Once the reason and motivation for the presentation have been explained, the shock presentation can then move on to introducing the presenter, again in a minimalist fashion. An example of this is shown in Figure 13.9.

With the shock presentation there is no need to beat around the bush. We quickly get to a discussion of the main ideas of the presentation, as shown in Figure 13.10.

While the presentation ordering used here is not necessarily the essential ordering of a shock presentation, in this example I decided to follow the main discussion with a more detailed and thorough motivation and consequence of the work, followed by the prior work slide

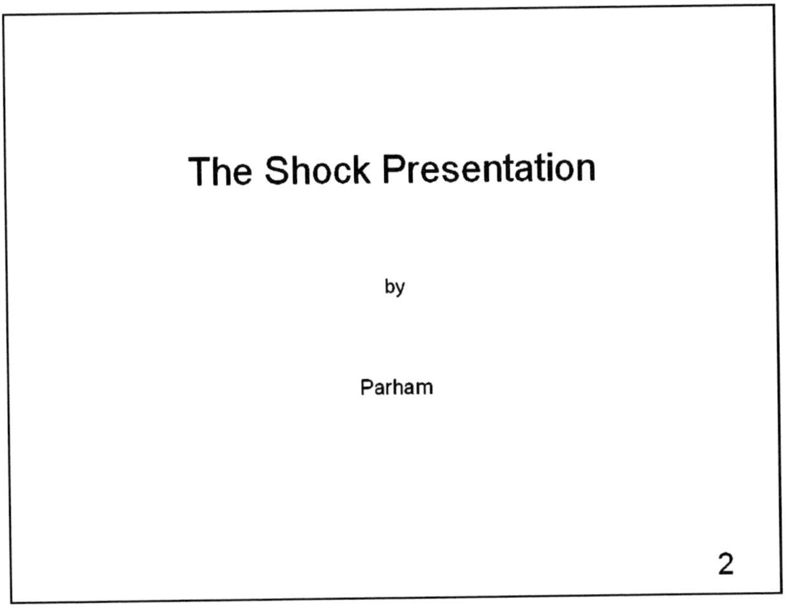

Figure 13.9. The shock presentation – Slide 2 – Title page.

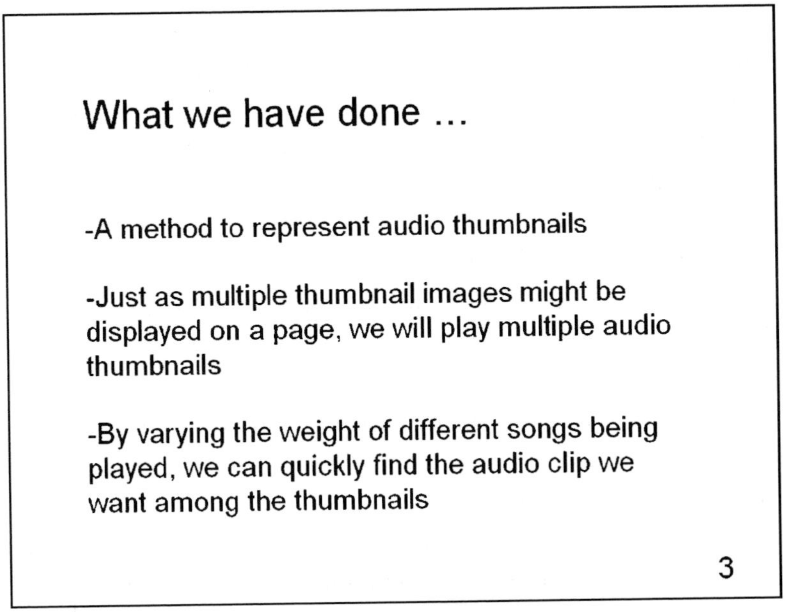

Figure 13.10. The shock presentation – Slide 3 – Discussion of the main ideas.

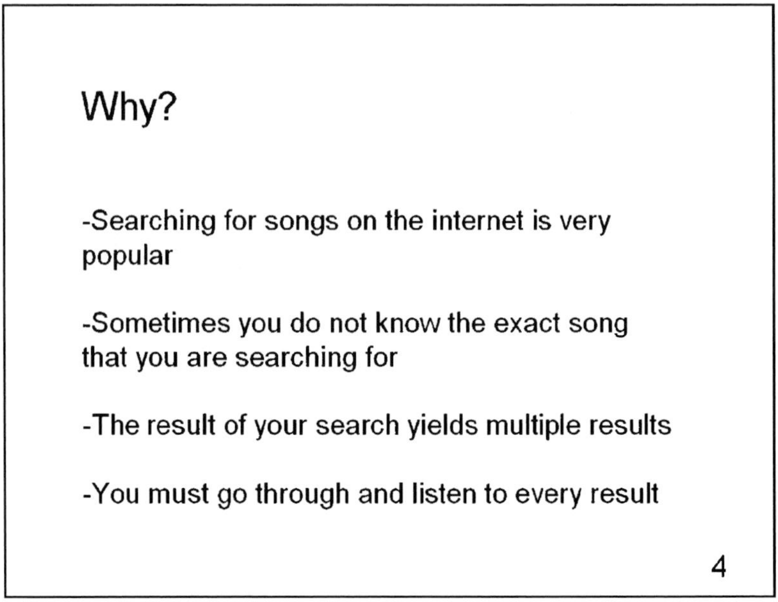

Figure 13.11. The shock presentation – Slide 4 – Motivation.

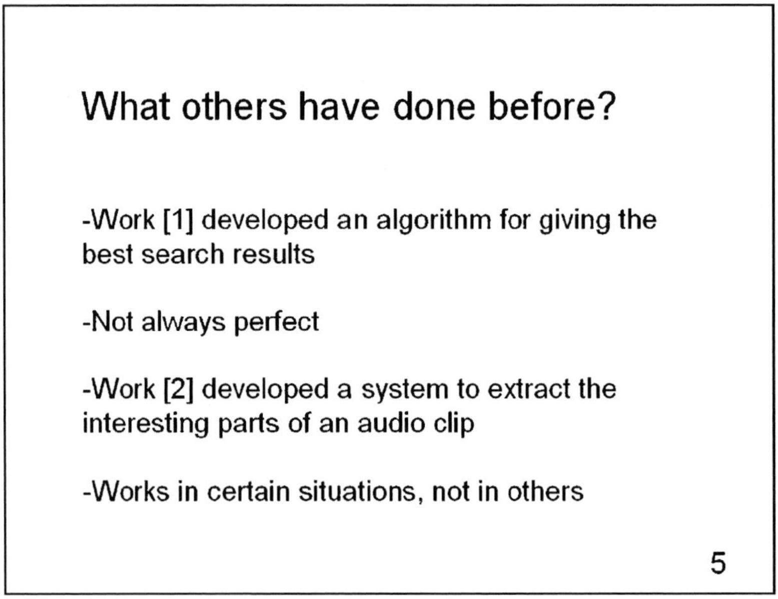

Figure 13.12. The shock presentation – Slide 5 – Prior work.

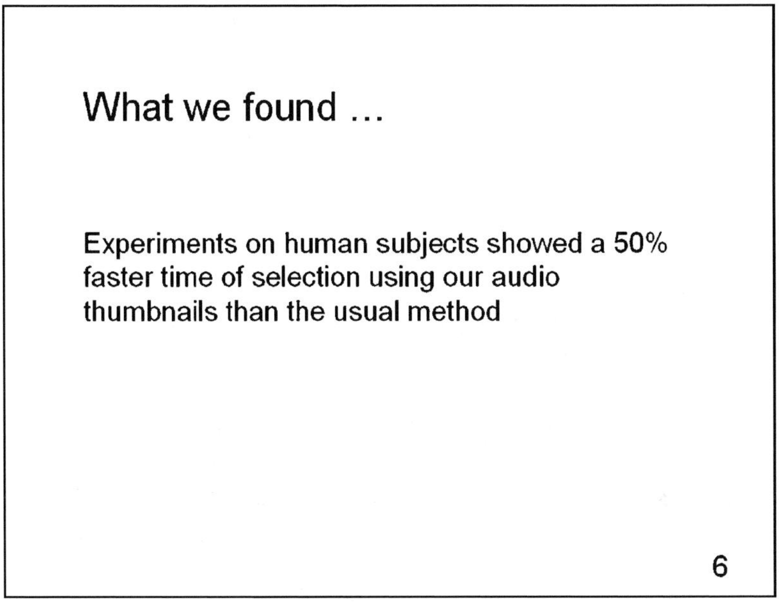

What we found ...

Experiments on human subjects showed a 50%
faster time of selection using our audio
thumbnails than the usual method

6

Figure 13.13. The shock presentation – Slide 6 – Summarized results.

and a summary of the overall results, as shown in Figures 13.11, 13.12, and 13.13.

The shock presentation is designed to be a natural and efficient method of presenting a topic without getting distracted by either the structure of the presentation or the traditional rules and constraints of presenting. If used effectively, it can result in a fantastic and memorable presentation. If used in a disorganized and irrational manner, it will result in a disastrous presentation. For this reason, it is perhaps advisable for novice presenters to start with the classic presenting style before attempting something that is a little bit more shocking.

13.3   A HYBRID APPROACH

Clearly both the classic presenting style and the shock presenting style are better in their own domains of application. For example, when making a career expanding or breaking presentation, a shock presentation might be a good bet, whereas presenting to an old fashioned corporate board, for example, might be best served with the selection of a safe and formal classic presentation. The only dilemma would be if you had to make a career expanding or breaking presentation to an old fashioned corporate board!

A third and better alternative is to combine elements of both presenting approaches to give a presentation that is natural, to the point, and exciting, while at the same time adhering to some of the formal definitions of a classic presentation. In the following case in point, we will analyze the anatomy of this "hybrid" style of presenting.

Case in point
The hybrid presentation is really a shock presentation repackaged and disguised as a classic presentation. It can start off very much like a classic presentation with a regular title page. However, it quickly focuses onto the main motivation of the presentation and the fundamental motivating question on which the presentation is based. These slides can be observed for our audio searching example in Figures 13.14, 13.15, and 13.16.

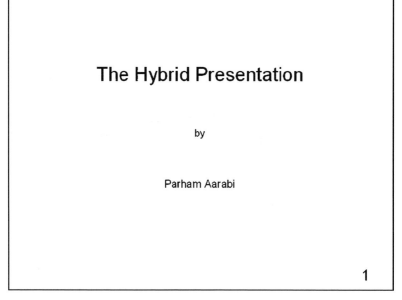

Figure 13.14. The hybrid presentation – Slide 1 – Title page.

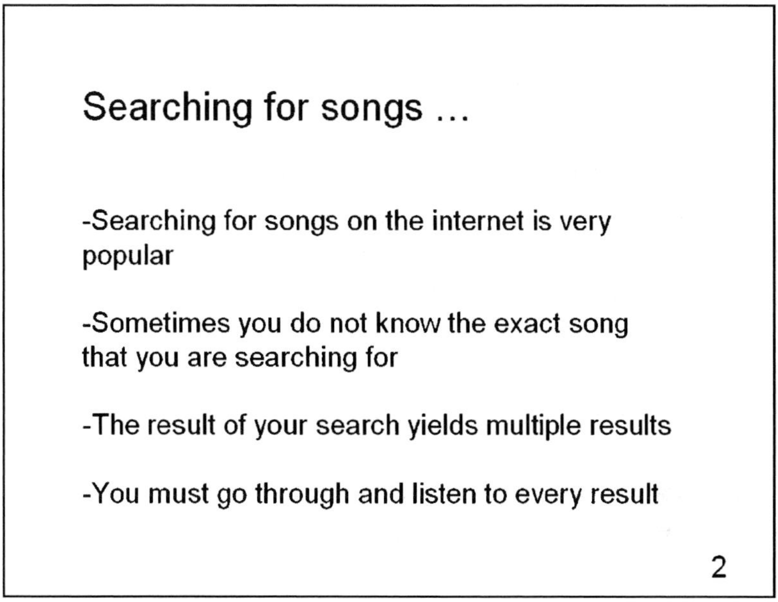

Figure 13.15. The hybrid presentation – Slide 2 – Motivation.

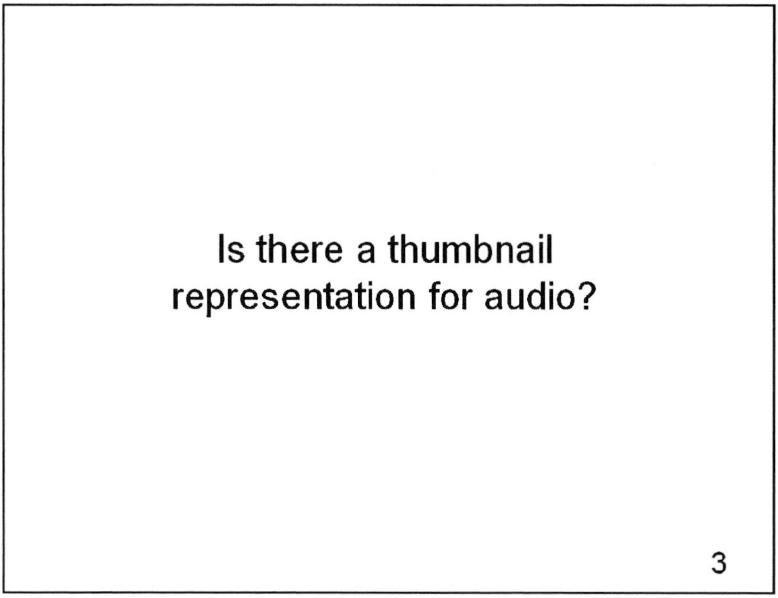

Figure 13.16. The hybrid presentation – Slide 3 – Motivating question.

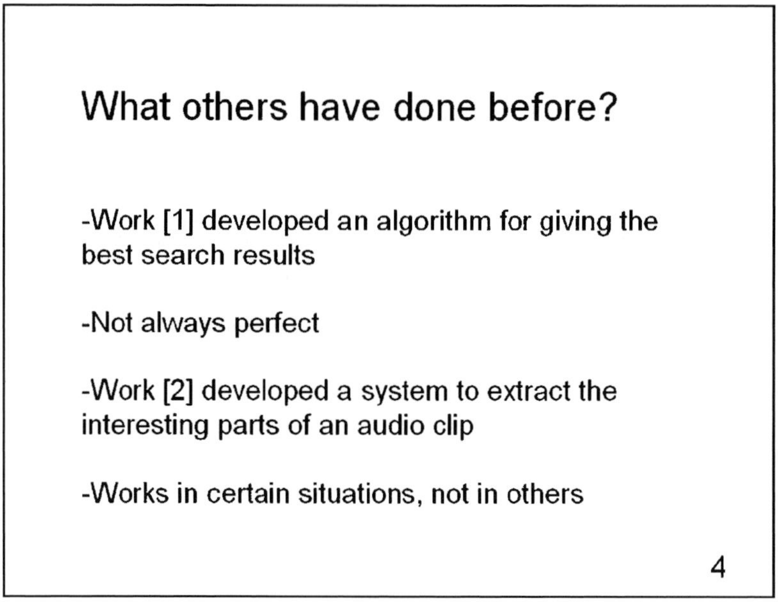

Figure 13.17. The hybrid presentation – Slide 4 – Prior work.

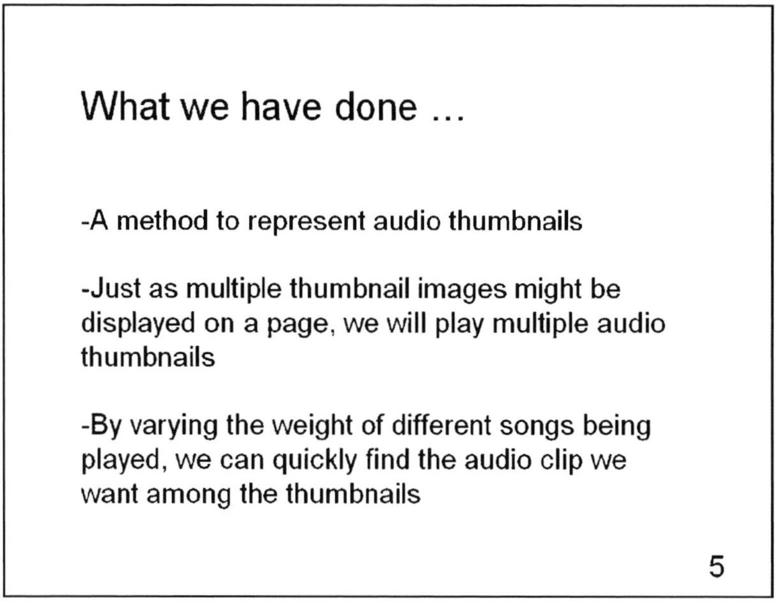

Figure 13.18. The hybrid presentation – Slide 5 – Discussion of the main ideas.

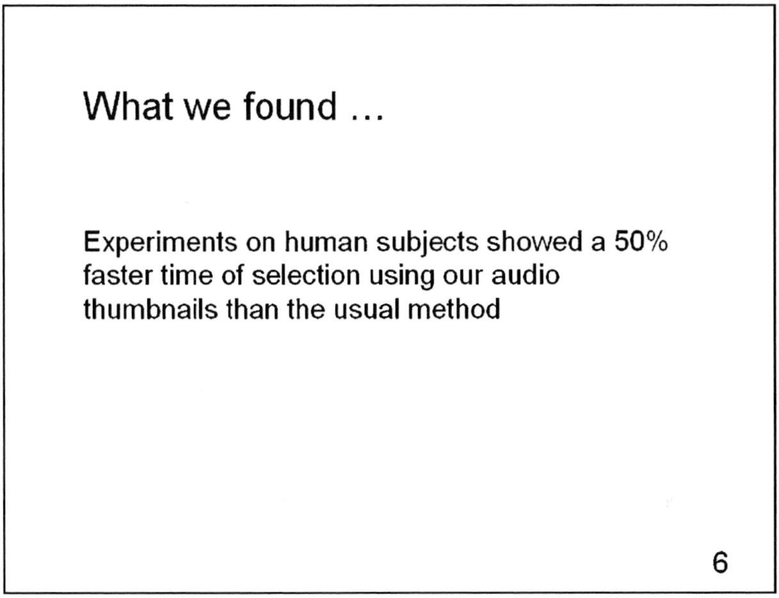

Figure 13.19. The hybrid presentation – Slide 6 – Summarized results.

From here, the hybrid presentation follows a natural flow starting from the prior work slide followed by the main ideas of the presentation and a summary of the results, as shown in Figures 13.17, 13.18, and 13.19.

While this ordering may be very similar to that of the classic presentation, the contents of the slides are most similar to a shock presentation. The language is informal, the tone is direct, and the focus is entirely on the main point of the presentation.

This hybrid presentation is the style that I most often recommend to my graduate students. It is not hard to master, yet with experience it can be used to deliver extremely memorable and exciting presentations all the while conforming to the regular rules of classic presenting which avoids raised eyebrows during thesis presentations or even presentations in front of an old fashioned corporate board!

13.4 CHAPTER SUMMARY

This chapter focused on the mechanics of professional presentations by studying three styles of presenting. The first style consisted of a traditional and classic presentation with a title page, outline slide, and so forth. This style of presenting, while very common and easy to master, can have limited effect in practice because of the rigidity of its structure. The alternative to this, the shock presenting style, has no structure or form and is just meant to shock and inspire the audience to listen to and to understand the presentation. While this can be very effective in certain settings, the lack of formality makes it non-ideal in some situations.

The final style of presenting that was mentioned was the hybrid approach. The hybrid presentation is as to-the-point and efficient as a shock presentation but with some minimal amount of conformation to the structure of the classic presentation. This is perhaps the best approach as it can be tailored to different situations where more or less formality, or, more or less shock, may be required.

13.5 Chapter checklist

- Professional presentations come in many different forms and varieties
- The most common are what we define as classic – they have a title page, an outline, introduction, motivation, prior work, main discussion, as well as ending with conclusions
- While this is good for a novice presenter, focusing too much on the structure of the presentation usually limits experienced presenters
- Another type of presentation is the shock style of presenting – here the goal is to shock and awe the audience by being direct and to the point regarding the main topics of the presentation
- Shock presentations can be funny, dramatic, or surprising – in all cases, the contents of the presentation, and not its form, are what matters
- Sometimes, a shock presentation may be inappropriate for a traditional audience

- A hybrid presentation is a combination of a shock presentation with a classic presentation – it obeys some of the rules of classic presentations while being, in essence, a shock-style presentation
- The hybrid presentation is perhaps the best choice for most audiences and for most presenters

14 Final words

After all that has been said, we get to this final chapter. By now, if you have read the book in detail, you know some of the strategies that can help make lectures successful and effective. However, there is much more to lecturing than what can be covered in any book. Lecturing requires persistence to keep on trying even after numerous failures. It requires practice and experience, just as practice and experience are required in order to become a competent artist. And finally, it requires keeping in mind why you are lecturing and why you are there. Just as a doctor is often faced with life and death questions regarding patients, so is a lecturer often faced with life altering questions regarding the audience. As a result, lecturers need to be active in what they do and how they lecture. They must avoid fear of trying new things and avoid getting stuck in lecturing local minima where their performance, while okay, could be substantially improved through further experiment and effort. In the next few sections, we will take a closer look at these and other general issues.

14.1 BE PERSISTENT

The first chapter of any success story is often about failure. The first time that you give a lecture, the first time that you try something new, do not expect immediate success. While you should learn from your mistakes, do not be disappointed by them. The difference between those who are successful and those who are not is that the successful people persist through failure by not giving up.

The above paragraph may seem like a collection of clichés pasted together, but they are all in fact true. When it comes to lecturing and dealing with large audiences, doing something wrong is inevitable. For example, it is possible that a test or exam may end up being far harder or easier than you imagined. Or, it may occur that in the middle of the lecture you get distracted and lose your train of thought. And of course, it may also be possible that you will give a lecture following which the entire audience shakes their heads in confusion and disappointment.

What will define you as a lecturer is not if these events occur, but rather how you respond to and counter such events. The best response is often to learn about and explore what went wrong, and then to take action to remedy the situation. If the lecture was confusing, give the entire lecture again. If you get distracted in the middle of a lecture, do not panic, just tell the audience that you forgot what you were going to say, take a little pause, and then continue on. And if you give a test or an exam that is too difficult, tell the students that and let them know about the fact that you will normalize the grades right then and there. To deal with lecturing errors, or any other errors for that matter, you must first see them, then you must learn from them, and finally, you must remain persistent in trying to resolve them and to remove their negative effect.

Case in point
As has been mentioned before, my initial attempts at lecturing were horrific disasters. My nervousness and inability as a lecturer would often result in unintelligible, unclear, and confusing lectures. Luckily, I was only a TA at Stanford when I started, giving me some time to fail in a safe environment prior to taking on my professorial position at the University of Toronto.

Also, since the courses at Stanford were televised live online and on the Stanford Instructional Television Network, I had the unique opportunity to view my lecturing mistakes over and over again. In time, once I had become comfortable with giving unclear and useless lectures, my nervousness and fear of failing disappeared. In other words, failing had become the norm for me. I realized that no matter what I did, it probably could not make things any worse. As a result, I began experimenting with the lectures by trying new methods and ways of lecturing. This experimentation, which allowed me to develop a sense of lecturing styles by the end of my Ph.D. at Stanford, continued when I started my career at the University of Toronto. The experimentation combined with my comfort with failing allowed me to learn from the experiments, get over any fears, and to give better lectures.

Usually, when someone wants to learn ice skating for the first time, it is common to first teach them about falling on the ice. Only by getting comfortable with falling, and seeing that you do not really die at the instant that you fall, can someone skate freely without being consumed by fear of failure. In a similar way, getting used to failing in lectures at the beginning of my career now allows me to lecture without the worry of failing. If I do make a mistake, then it just becomes a matter of "been there, done that," which allows me to focus on correcting the mistake rather than being paralyzed by it.

14.2 DON'T FORCE THINGS THAT ARE NOT MEANT TO BE

While persistence is generally a good thing, there is a clear difference between intelligent and adaptive persistence, where you push for a certain outcome by adjusting and learning from errors, and blind and brute persistence, which is an unintelligent push forward without any care or regard for the consequences. The latter form of persistence is almost universally a bad thing, from lectures, to careers, to personal lives.

As a lecturer, there will often come a time when you try something new and it will not work. Perhaps you try to explain a scientific principle to the audience in a unique way and the audience becomes confused, or you try an example which, while seeming clear to you, is useless if not confusing to the audience. In such cases, it is important to keep in mind that not everything you do is required to be successful. We all make errors; errors which initially may have seemed to have the potential for significant success. When these errors occur, it is very important to learn from how and why they occurred. It is almost always a mistake to keep trying the same thing over and over again, after repeated failures, without some form of adaptation or learning.

This is especially true when dealing with people. Situations such as lectures require constant adaptation and revision of strategies. If you present a topic in a certain way that most of the audience do not understand, or give an assignment/quiz/test whose difficulty is significantly mismatched with the abilities of the audience, then problems will arise. It is important to be aware of the fact that if you force the audience to solve harder assignments or write harder tests, it does not necessarily make them smarter. Or if you decide to teach all the advanced topics instead of the basic topics, your audience will likely not be better off.

Pushing the audience to the outer fringes of what they can do is a good thing. It is only by this push that they will learn and grow. However, pushing them beyond what they can accomplish will disappoint them and replace their interest in the topic with fear. It is tragic sometimes to see lecturers who will make any excuse for a bad lecture or a bad course, including blaming the audience for lack of intelligence or blaming the textbook for lack of clarity. Sometimes, the only blame that can be assigned must be placed on the lecturer for simply pushing too hard with a flawed lecturing plan and strategy.

14.3 MORE ART THAN SCIENCE

There are things in life that can be fully modeled by mathematics and scientific principles. For these things, people rightfully study the details of the principles in order to gain a deeper and better understanding of the topic. On the other hand, there are certain things in life that are inherently artistic. Some may place mathematical rules and scientific principles on them, but such rules and principles are often inadequate for deeper insight and explanation.

Of course, there is the final class of things which are both scientific and artistic. Lecturing is one such example. There are aspects of lectures that can be quantified, such as what the general elements of a successful lecture are, or, what the capacity of the audience is and how it changes with respect to time. These quantifiable elements have been covered in this book and a variety of other books on public speaking. However, it should always be kept in mind that lecturing is as much an art as it is a science, if not more so. As a result, just as the best way to become a master painter is to paint, instead of reading about painting, or the best way to dance is to actually do it, rather than to read about it, the best way to learn about lecturing is by getting experience in front of an audience and letting your inner feelings and intuition run the show.

Case in point

When giving a lecture, try to look the audience directly in their eyes. Try feeling their emotions, their fears, their hopes, and their thoughts. I have during the course of my life had the pleasure of painting, playing the piano, and experiencing other art forms. The issues and emotions of lecturing are in fact no different. For example, the feeling you get when you pick up a brush to paint is in many ways similar to when you pick up the chalk to start your lecture. No one can tell you exactly how and what you should paint, and in a fairly similar way, no one can or should tell you how you should lecture. All that people can say is whether in the end, your painting, or your lecture, was successful or not.

When I first pace the lecture room prior to the start of a lecture, and stand in front of hundreds of students, I can almost sense or feel their energy. This reminds me of when I was a teenager studying karate. In those karate classes, I was taught not only to look at my opponents and others in the room, but to sense and feel them. This sensation and feeling utilizes all available senses for an integrated view of the environment and its occupants. This view, which enhances the speed, vision, and response time of a martial artist, can be just as effective for a lecturer by increasing their presence and vision throughout the lecture room. This heightened awareness can be a key advantage to a lecturer for detecting random conversations, audience fatigue, as well as other information and trends among the audience.

14.4 MOST IMPORTANT OF ALL ...

In nearly 30 years of life, there are a few things that I have learned, and a great many more things that I still need to learn. Life is filled with challenges and obstacles, with people who continuously say that you can't do this and you won't achieve that. Even worse than these people are the limits and constraints that we place on ourselves. Have you ever been interested in a girl that you always daydreamed about, but were always too scared and fearful to talk to? Have you ever stopped going after an opportunity because you feared failure? Or, have you ever given a lecture that was so bland because you feared trying something new, something different, something risky?

Perhaps you have, and perhaps you have not. I personally have often experienced these situations in my life, and have been frustrated and angry after each incident. There has always been something in me, perhaps something genetic, perhaps a deeply ingrained shyness, which I have had to fight continuously; something that has tried to stop me from saying what I felt and grabbing opportunities that I believed in. This fear, shyness, and hesitation exists in every person to a greater or lesser extent and in different ways, and for some it is undetectable. The worst thing for us to do is to give up in the face of such fears. Or, for example, to believe as fact that what we fear going after, is either something we do not want or something we cannot have. I have lost several opportunities because I feared failure. But slowly, as the number of lost opportunities piled up and the personal resentment and anger in me grew, I realized that this hesitation, which manifests itself differently in different situations, is really the greatest challenge for success in my life. As a result, I consistently pushed myself to grab opportunities, to talk to those around whom I would normally be shy, and to give the best darn lectures that my body, mind, voice, and throat could tolerate. After 30 years, the most important thing of all that I have learned is to put 100% effort into those valuable and once-in-a-lifetime opportunities, to try your hardest, to prepare your best, to think and clarify your thoughts to the greatest extent possible, and when it comes to lectures, to put on one heck of a show that you (and hopefully the audience) will never forget!

Thank you for reading this book.

Parham Aarabi

About the Author

Parham Aarabi is a Canada Research Chair in Internet Video, Audio, and Image Search, an Associate Professor in The Edward S. Rogers Sr. Department of Electrical and Computer Engineering at the University of Toronto, and the founder and director of the Artificial Perception Laboratory.

He received his Ph.D. (2001) in Electrical Engineering from Stanford University, his M.A.Sc. (1999) in Computer Engineering from the University of Toronto, and his B.A.Sc. (1998) in Engineering Science (Electrical Option) from the University of Toronto. In 2001, he served as a co-instructor in the Electrical Engineering Department at Stanford University. He has also served as a teaching assistant for a variety of courses at the University of Toronto and Stanford University from 1999 to 2001.

In 2001, at the age of 24, Parham Aarabi became one of the youngest faculty members in the University of Toronto, teaching courses on digital systems design and probability theory. In 2002, he was selected by the Electrical and Computer Engineering students to receive the Best Computer Engineering Professor award. In 2003, he was again selected to receive the same, but renamed, Professor of the Year award. In 2004, he received the same (but yet again renamed) Departmental Teaching award.

In 2003 he was selected by the Faculty of Engineering to receive the Early Career Teaching award, given in recognition of his "superb accomplishment in teaching." In 2004, Parham Aarabi was selected to receive the inaugural IEEE Mac Van Valkenburg Early Career Teaching Award. This prestigious award, which has since 2004 been given on an annual basis to a single IEEE Education Society member world-wide, was awarded based on Parham Aarabi's "outstanding contributions to electrical and computer engineering education, including exemplary classroom teaching and inspirational mentoring of undergraduate students in research projects."

In 2005, he was awarded the Gordon Slemon Teaching of Design Award, given in recognition of his supervision of the 2005 ECE-APL Robotics Competition. Also in 2005, and subsequently in 2006, he was selected by TVO as one of the top lecturers among all fields in the province of Ontario (one of only two engineering professors in 2005 to be included in this list). Finally, in the fall of 2005, Parham Aarabi was selected by MIT's *Technology Review* as one of the "world's top young innovators," known as the TR35 award. The winners of this award consist of the top 35 innovators in the world under the age of 35.

Parham Aarabi's research has focused on the interface between humans and computers. His work in this area has appeared in more than 60 publications and has been covered by media such as the *New York Times*, MIT's *Technology Review* magazine, *Scientific American*, *Popular Mechanics*, the Discovery Channel, CBC Newsworld, Tech TV, Space TV, and City TV.

In recognition of these teaching and research achievements, Parham Aarabi was awarded tenure in 2005 at the age of 28, becoming one of the youngest tenured faculty members at the University of Toronto. The following year, he received the provincial Early Researcher Award (formerly known as the Premier's Research Excellence Award). Also, in recognition of his lecturing and teaching achievements, he was awarded the 2006 Student's Administrative Council and Association of Part-time Undergraduate Students' University-wide Undergraduate Teaching Award.

Index